COAL
COMBUSTION

COAL COMBUSTION

by Jerzy Tomeczek, Ph.D., D.Sc.
Professor of Fuel Engineering
Silesian Technical University

KRIEGER PUBLISHING COMPANY
MALABAR, FLORIDA
1994

Original Edition 1994

Printed and Published by
KRIEGER PUBLISHING COMPANY
KRIEGER DRIVE
MALABAR, FLORIDA 32950

FROM A DECLARATION OF PRINCIPLES JOINTLY ADOPTED BY A COMMITTEE OF THE
AMERICAN BAR ASSOCIATION AND A COMMITTEE OF PUBLISHERS:

This publication is designed to provide accurate and authoritative information in regard to the subject
matter covered. It is sold with the understanding that the publisher is not engaged in rendering legal,
accounting, or other professional service. If legal advice or other expert assistance is required, the services
of a competent professional person should be sought.

Library of Congress Cataloging-In-Publication Data

Tomeczek, Jerzy.
   Coal combustion / by Jerzy Tomeczek.
     p.   cm.
   Includes bibliographical references.
   ISBN 0-89464-651-6
    1. Coal—Combustion. I. Title.
  TP325.T66 1994
  662.6'2—dc20
                                     91-46266
                                       CIP

10 9 8 7 6 5 4 3 2

CONTENTS

Contents

NOMENCLATURE

A — ash content in coal

A_B — bed cross-section area, m^2

A_{B1} — bed cross-section area per one distributor port, m^2

A_o — cross-section area of a single distributor port, m^2

A_p — particle surface area, m^2

Ar — Archimedes number

A_x — internal surface area of porous particle, m^2

b_i — burn out of i-th fraction of coal

C — carbon content in coal

C_b — gas concentration in bubbles, $kmol/m^3$

C_e — gas concentration in emulsion, $kmol/m^3$

C_{CO} — concentration of carbon monoxide, $kmol/m^3$

C_{O_2} — concentration of oxygen, $kmol/m^3$

C_{O_2,d_p} — concentration of oxygen at particle surface, $kmol/m^3$

$C_{O_2,e}$ — concentration of oxygen in emulsion, $kmol/m^3$

$C_{O_2,o}$ — concentration of oxygen at bed inlet, $kmol/m^3$

$C_{O_2,\infty}$ — concentration of oxygen at large distance from the surface, $kmol/m^3$

C_o — concentration at nozzle outflow, $kmol/m^3$

$[C_f]$ — surface concentration of free active centres, $centres/m^2$

$[C_t]$ — surface concentration of total active centres, $centres/m^2$

$[C(O)]$ — surface concentration of occupied active centres, $centres/m^2$

Cl — cloride content in coal

c — specific heat, $J/(kg \circ K)$

c_a — specific heat of ash, $J/(kg \circ K)$

c_o — specific heat of organic matter, $J/(kg \circ K)$

c_p — specific heat of particle, $J/(kg \circ K)$

c_{pg} — specific heat of gas at constant pressure, $J/(kg \circ K)$

D — diffusion coefficient, m^2/s

D_{ij} — diffusion coefficient of i-th component through j-th component, m^2/s

D_{Kj} — Knudsen diffusion coefficient of j-th component, m^2/s

D_{O_2} — oxygen diffusion coefficient, m^2/s

d_B — bed diameter, m

d_b — bubble diameter, m

d_{bo} — initial bubble diameter, m

d_{bm} — maximum bubble diameter, m

d_e — effective nozzle diameter, m

d_f — flame sphere diameter, m

d_n — distributor nozzle diameter, m

d_o — nozzle diameter, m

d_p — particle diameter, m

$d_{p,i}$ — diameter of i-th group of particles, m

d_{po} — initial particle diameter, m

$d_{po,i}$ — initial particle diameter of i-th fraction of coal, m

d_t — tube diameter, m

d_x — pore diameter, m

E — activation energy, J/kmol

Fr — Froude number

G — momentum, N

G_o — momentum of gas at nozzle outflow, N

G_x — axial momentum, N

G_φ — angular momentum, Nm

g — gravity, m^2/s

g_i — mass fraction of i-th component

$g_{o,i}$ — initial mass fraction of i-th group of particles

H — hydrogen content in coal

H_l — heat of combustion (lower), J/kg or J/kmol

$H_{u,n}$ — heat of combustion (upper) at standard state (p_n, T_n), J/kg or J/kmol

ΔH_d — enthalpy of coal decomposition, J/kg

h — height, m

h_B — bed height, m

$h_{B,o}$ — stagnant bed height, m

$h_{B,mf}$ — bed height at minimum fluidization velocity, m

h_{bo} — height of vertical jet penetration above perforated distributor, m

h_e — height of particle entrainment, m

h_o — nozzle width of plane burner, m

i — specific enthalpy, J/kg

i_l — specific enthalpy of l-th component, J/kg

k — rate constant of effective combustion reaction, m/s

k_b — rate constant of backward reaction, $\dfrac{1}{s}\left(\dfrac{kmol}{m^3}\right)^{-1}$

k_C — rate constant of chemical reaction, m/s

k_f — rate constant of forward reaction, $\dfrac{1}{s}\left(\dfrac{kmol}{m^3}\right)^{-1}$

k_j — rate constant of j-th surface elementary reaction, $s^{-1}\,(kmol/m^3)^{-1}$ or s^{-1}

k_m — reaction rate constant, $\dfrac{kmol}{s\,kg}\left(\dfrac{kmol}{m^3}\right)^{-1}$ or $\dfrac{1}{s}\left(\dfrac{kmol}{m^3}\right)^{-1}$

k_o — preexponential factor, m/s or kg/(m^2 ∘ s ∘ Pa)

Le — Lewis number

L_f — flame length, m

L_g — mean free path of diffusing gas, m

L_k — combustion chamber width (or diameter), m

L_s — depth of jet penetration into a cross flowing gas, m

L_x — pore length, m

M — molecular mass, kg/kmol

M_C — carbon molecular mass, kg/kmol

M_{CO} — molecular mass of CO, kJ/kmol

M_j — molecular mass of j-th component, kg/kmol

m — mass, kg

m_B — mass of particles in bed, kg

m_{Bi} — mass of i-th group of particles in bed, kg

m_C — mass of carbon atom, kg/carbon centre

m_l — mass of liquids released from coal, kg/kg coal(waf)

m_p — particle mass, kg

m_V — mass of volatiles released from coal, kg/(kg coal(waf))

m_∞ — mass of volatiles released from coal after very long residence time $t \rightarrow \infty$, kg/(kg coal (waf))

$\dot{m}$ — mass flow, kg/s

$\dot{m}_a$ — mass rate of fines generated as result of attrition, kg/s

$\dot{m}_e$ — mass flow of elutriated particles, kg/s

$\dot{m}_g$ — mass flow of gas, kg/s

$\dot{m}_{go}$ — mass flow of gas at nozzle outflow, kg/s

$\dot{m}_o$ — mass flow at nozzle outflow, kg/s

$\dot{m}_{ov}$ — mass flow through overflow, kg/s

$\dot{m}_{po}$ — mass flow of particles at nozzle outflow, kg/s

$\dot{m}_{po,i}$ — mass flow of i-th group of particles at nozzle outflow, kg/s

$\dot{m}_r$ — mass flow of recirculating gas, kg/s

$\dot{m}_x$ — mass flow in x direction, kg/s

N — nitrogen content in coal

N_j — order of reaction of j-th product formation

N_x — number of pore mouths per unit mass of particle, pores/kg

n_b — frequency of bubbles rising, 1/s

$\dot{n}$ — molar flow, kmol/s

O — oxygen content in coal

Pr — Prandtl number

p — pressure, Pa

p_1 — pressure within 1μm pores, Pa

p_6 — pressure within 6μm pores, Pa

p_n — standard pressure, Pa

p_{O_2} — oxygen partial pressure, Pa

p_{O_2,d_p} — oxygen partial pressure at particle surface, Pa

Δp_B — pressure drop in bed, Pa

Δp_d — pressure drop at distributor, Pa

Q — heat, J

Q_{ch} — heat of chemical reactions, J/m^3

$\dot{Q}_G$ — rate of heat generation, W

$\dot{Q}_L$ — rate of heat losses, W

$\dot{q}$ — surface heat flux, W/m^2

$\dot{q}_r$ — radiation heat source, W/m^3

R — gas constant, J/(kmol ∘ K)

R_f — fouling index

R_s — slagging index

$\dot{R}$ — rate of reaction, kmol/s or kg/s

$\dot{R}_C$ — rate of chemical reaction, kmol/s

$\dot{R}_D$ — rate of oxygen diffusion to the particle surface, kmol/s

$\dot{R}_i$ — rate of i-th homogeneous reaction, kmol/(s ∘ m³) or kg/(s ∘ m³)

$[\dot{R}]$ — rate of surface reaction, kg/(s ∘ m²)

Re — Reynolds number

Re_{mf} — Reynolds number at minimum fluidization velocity

r — geometrical coordinate (radius), m

S — sulphur content in coal

S — swirling number

S_{pyr} — pyritic sulphur in coal

S_t — total sulphur in coal

Sc — Schmidt number

Sh — Sherwood number

s — distance between rows of vertical tubes, m

T — temperature, K or °C

$T_{1/2}$ — hemispherical temperature, K or °C

T_{25} — temperature of ash viscosity equal 25(Pa ∘ s), K or °C

T_a — temperature of gas surrounding the jet, K or °C

T_e — temperature of emulsion, K or °C

T_f — temperature of jet flame, K or °C

T_g — temperature of gas, K or °C

T_{cv} — temperature of critical viscosity, K or °C

T_n — standard temperature, K or ° C

T_o — temperature of gas at nozzle outflow, K or °C

T_q — temperature of cross flowing gas, K or °C

T_p — particle temperature, K or °C

T_{pi} — particle ignition temperature, K or °C

$T_{p,i}$ — temperature of i-th group of particles, K or °C

T_w — temperature of wall, K or °C

Th — Thiel number

t — time, s

t_C — particle-wall contact time, s

t_c — time of combustion, s

t_V — time of devolatilization, s

V — volatile matter in coal

V_b — bubble volume, m³

V_p — particle volume, m³

V_x — internal porous volume of particle, m³

$\dot{V}$ — volume flow, m³/s

w — velocity, m/s

w_a — absolute bubble velocity, m/s

w_b — velocity of isolated bubble in stagnant liquid, m/s

w_g — gas velocity, m/s

w_n — velocity at distributor nozzle, m/s

w_o — velocity at nozzle outflow, m/s

w_{opt} — velocity of maximum heat transfer coefficient, m/s

w_q — velocity of cross flowing gas, m/s

w_x — axial component of velocity, m/s

$w_{6,V}$ — velocity of volatiles within $6\mu m$ pores, m/s

w_t — terminal fluidization velocity, m/s

w_φ — tangential velocity, m/s

x — geometrical coordinate (axial coordinate), m

y — geometrical coordinate (perpendicular to x), m

z — geometrical coordinate, m

z_{CO} — molar fraction of CO in gas

z_{CO_2} — molar fraction of CO_2 in gas

α — heat transfer coefficient, $W/(m^2 \circ K)$

α_g — gas convection heat transfer coefficient, $W/(m^2 \circ K)$

α_p — particle convection heat transfer coefficient, $W/(m^2 \circ K)$

α_r — radiation heat transfer coefficient, $W/(m^2 \circ K)$

β — mass transfer coefficient, m/s

β_{bc} — bubble-cloud mass transfer coefficient, m/s

β_{be} — bubble-emulsion mass transfer coefficient, m/s

β_{ce} — cloud-emulsion mass transfer coefficient, m/s

ε — bed porosity

ε_b — volume fraction of bubbles in bed

ε_{mf} — bed porosity at minimum fluidization velocity

ε_o — stagnant bed porosity

ε_{p-w} — particle-wall radiation exchange factor

ε_w — wall radiation emissivity

η — dynamic viscosity, $Pa \circ s$

η_{ef} — effective dynamic viscosity, $Pa \circ s$

η_g — dynamic viscosity of gas, $Pa \circ s$

η_t — turbulent dynamic viscosity, $Pa \circ s$

Θ — Thring-Newby parameter

Θ_1 — volume share of $1\mu m$ pores

Θ_6 — volume share of $6\mu m$ pores

Θ_d — distributor porosity

Θ_s — outside particle surface porosity

λ — thermal conductivity, $W/(m \circ K)$

λ_d — drag number

λ_e — thermal conductivity of emulsion, $W/(m \circ K)$

λ_e^o — thermal conductivity of packed bed, $W/(m \circ K)$

λ_f — friction number

λ_g — thermal conductivity of gas, $W/(m \circ K)$

λ_p — particle thermal conductivity, $W/(m \circ K)$

ν — kinematic viscosity, m^2/s

ν_g — kinematic viscosity of gas, m^2/s

$\nu_{6,\,v}$ — kinematic viscosity of volatiles within 6μm pores, m^2/s

ξ_d — distributor discharge number

ρ — density, kg/m^3

$\rho_{6,v}$ — density of volatiles within 6μm pores, kg/m^3

ρ_a — density of gas surrounding the jet, kg/m^3

ρ_g — gas density, kg/m^3

ρ_{go} — initial gas density, kg/m^3

ρ_o — density at nozzle outflow, kg/m^3

ρ_p — particle density, kg/m^3

ρ_q — density of cross flowing gas, kg/m^3

ρ_{po} — initial particle density, kg/m^3

σ — Stefan-Boltzmann constant, W/(m$^2 \circ$ K^4)

τ_6 — tortuosity of 6μm pores

φ — particle shape factor

φ_C — half angle of jet concentration profile, degree

φ_w — half angle of jet velocity profile, degree

ω — angular velocity, 1/s

1
COAL PROPERTIES

Coal is a fossil substance formed mainly as a result of temperature and pressure action on the fallen remains of plants. The properties of coal are determined by both the physical parameters during the stages of coal formation and the type of origin plants in each geological era. The organic part of coal is usually accompanied by a certain amount of water and inorganic compounds. The content of carbon in coal defines the degree of carbonification which does increase along the row of substances:

Peat → Lignite → Bituminous Coal → Anthracite.

The differences between the origin plants and the extent of their decomposition during the first stage of coal formation determine the petrographic components called macerals, while the pressure and temperature during the following geochemical stages cause differences in carbonification degree. Macerals, called so by analogy to minerals, can be distinguished only by a microscope. The difference between macerals diminishes with increasing degrees of carbonification. Macerals are the most elementary uniform petrographic formations differentiated from each other by chemical and physical properties. There are two ways of petrographic analysis:

- transmitted light method,
- reflected light method.

It has to be stated clearly, that only the petrography of hard coal is well-ordered, so speaking about the petrographic entities we mean always hard coal. The macerals are most frequently divided into three groups:

Vitrinite — Macerals of this group have a clear fibrous cellular structure of the origin woody tissues. Reflectance of vitrinite increases with the degree of carbonification. In reflected light it appears from gray to light gray or white. Vitrinite has good coking properties (high agglomerating properties) and poor mechanical properties.

Liptinite — Macerals of this group originate from plant substances other than woody tissues. Reflectance of a polished sample is lower than of vitrinite. In reflected light it appears from dark gray to light gray. Liptinite has higher hydrogen content than vitrinite. During heating it transforms into plastic melt which decomposes forming large amount of tar. Liptinite has good coking and good mechanical properties.

Inertinite — These macerals are completely structureless. Reflectance of inertinite is higher than of vitrinite. In reflected light it appears from light gray to white or yellowish white. During heating inertinite behaves like an inert material. Mechanical properties are poor.

Figure 1.1. presents a simplified scheme of hard coal petrography based mainly on data by Stach et al. (1982). Macerals as uniform formations create microlithotypes which can be also identified only by a microscope. The only

1

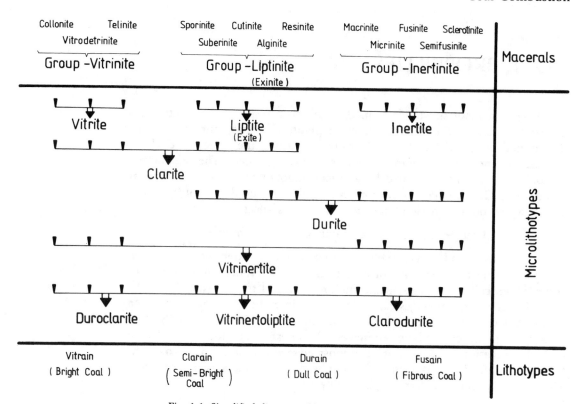

Fig. 1.1. Simplified diagram of hard coal petrography.

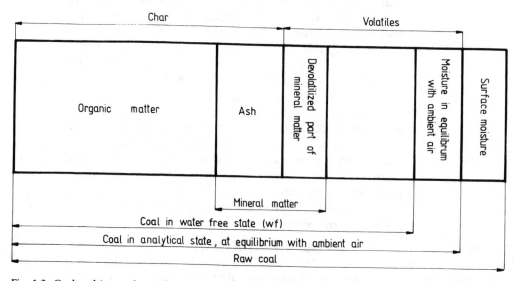

Fig. 1.2. Coal and its products after heating in inert atmosphere at standard temperature (850°C, 900°C).

macroscopic recognizable structures are the lithotypes, which are formed of microlithotypes.

Because properties of coal samples can vary with time and temperature, it is then necessary to declare the state of coals in order to describe them. In practice the following four states are commonly used:

"a" — Analytical state in which coal is in thermodynamic equilibrium with ambient air (moisture in coal in equilibrium with air moisture).

"d" or "wf" — Dry state (water free) which can be obtained after 2h residence of the coal sample in a furnace of temperature 105°C–110°C.

"waf" — Water and ash free state.

"af" — Ash free state.

During heating both the organic and mineral matter of coal decompose. Heated in an inert atmosphere coal releases volatile substances: water, steam, volatile tar, and gases. The first two components of volatiles condense at ambient temperature while the third does not, so often a terminology is used:

- liquid volatiles,
- gaseous volatiles.

The remaining solid body containing organic substance and ash is called char or coke. Because the extent of decomposition depends mainly on the temperature, it is necessary to define the standard temperature conditions. Figure 1.2. presents the characteristics of coal and its products after heating in an inert atmosphere at standard temperature (850°C–900°C). If heating took place in air then the combustible products would burn out.

For practical purposes the following three kinds of analysis are used to characterize coals:

- proximate analysis → moisture, ash, volatile matter,
- ultimate analysis → content of elements: C, H, S, N and O,
- miscellaneous analysis → heating value, forms of sulphur, chlorine, trace metals, CO_2, free swelling index or Roga index, dilatation, ash fusibility, ash composition.

1.1. COAL CLASSIFICATION

The international system of hard coal classification presented in table 1.1. is based on a three digit code:

1st digit → 0–9 → Class,
2nd digit → 0–3 → Group,
3rd digit → 0–5 → Subgroup.

This system utilizes the following four features of coal determined in standarized conditions:

Class — Volatile content in "waf" state = Mass of coal which is released in form of volatiles after a few minutes residence at standard temperature (850°C–950°C) in an inert atmosphere.

Table 1.1 INTERNATIONAL CLASSIFICATION OF HARD COALS

First figure of the code number indicates the class of coal, determined by volatile matter content up to 33% and by calorific parameter above 33%.
Second figure indicates the group of coal, determined by caking properties.
Third figure indicates the sub-group, determined by coking properties.

Group Number	Free Swelling Index	Roga Index	Class 0	Class 1	Class 2	Class 3	Class 4	Class 5	Class 6	Class 7	Class 8	Class 9	Sub-group number	Dilatometer Test (% dilat.)	Gray-King Assay
3	>4	>45					435	535	635				5	>140	>G8
						334	434	534	634				4	50-140	G5-G8
						333	433	533	633	733			3	0-50	G1-G4
						332a | 332b	432	532	632	732	832		2	<0	E-G
2	2.5-4	20-45				323	423	523	623	723	823		3	0-50	G1-G4
						322	422	522	622	722	822		2	<0	E-G
						321	421	521	621	721	821		1	contraction only	B-D
1	1-2	5-20			212	312	412	512	612	712	812		2	<0	E-G
					211	311	411	511	611	711	811		1	contraction only	B-D
0	0-0.5	0-5		100 (A | B)	200	300	400	500	600	700	800	900	0	non-softening	A

Class number	0	1	2	3	4	5	6	7	8	9
Volatile matter (waf)	0-3	>3-10 (>3-6.5 | >6.5-10)	>10-14	>14-20 (>14-16 | >16-20)	>20-28	>28-33	>33	>33	>33	>33
Heat of combust. kJ/kg (af) (equil. with air 30°C, 96% hum.)	–	–	–	–	–	–	>32450	>30150-32450	>25540-30150	≤25540

Classes have an approximate volatile matter content:
6: 33-41 %
7: 33-44 %
8: 35-50 %
9: 42-50 %
coal samples ≤ 10% ash

Group — Caking properties = Characterized by alternative parameters: free swelling index or Roga index, determining the mechanical endurance of solid sample after devolatilization.

Subgroup — Coking properties = Characterized by alternative parameters: Gray-King coke type or dilatation, determining the relative size change of formed coal pistil heated with 5 K/min rate at constant pressure in an inert atmosphere.

The fourth parameter in table 1.1. indicates the heat of combustion of coal at ash free state containing moisture at equilibrium with air of 30°C temperature and 96–97% humidity. Heat of combustion is used as the basis for class determination for coals with volatile content higher than 33% (waf).

The international system of hard coal classification introduced in 1956 did not eliminate the traditional systems used in various countries. Table 1.2. presents a simplified comparison of American, German and Polish systems in which coals are divided into types on the basis of generally the same parameters as in the international system. First of all it should be noticed that the borders of coal types and of classes do not correspond exactly with each other. Second, the table demonstrates the shortcomings of the coal types, some of them spread over many classes, that justifies the demand of common use of the international system.

Lignite classification is not as developed as the hard coal classification. Nevertheless, for coal with heat of combustion below 23870 kJ/kg (af) con-

Table 1.2 **Comparison of national systems of hard coal classification according to types.**

Intern. Class	USA	Germany	Poland						
0	Meta-anthracite	Antrazit	Metaantracyt						
1	Anthracite		Antracyt						
2	Semi-anthracite	Mager Kohle	Antracytowy						
3	Low volatile bituminous	Ess Kohle	Chudy	Semikoksowy	Ortokoksowy	Meta-koksowy			
4	Medium volatile bituminous	Fett Kohle				Gazowokoksowy		Gazowo-plomienny	Plomienny
5	High volatile bituminous A	Gas Kohle					Gazowy		
6									
7	High volatile bituminous B	Gas Flammkohle							
8									
9	High volatile bituminous C Sub-bituminous A, B, C								

taining moisture at equilibrium with air of 96–97% humidity, a system of classification based on two parameters is proposed:

Class → 1–6, according to moisture content in raw coal (af),

Group → 0–4, according to tar generation in % of (waf) state coal

1.2. COAL ASH

The mineral matter in coal is almost unbound chemically with the organic matter. The main components of the mineral substance are (Laurendeau, 1979):

clay: kaolinite $Al_2Si_2O_5(OH)_4$, illite $KAl_3Si_3O_{10}(OH)_2$ – 50% mass,
carbonates: calcite $CaCO_3$, siderite $FeCO_3$, dolomite $CaCO_3 \cdot MgCO_3$, ankerite $2CaCO_3 \cdot MgCO_3 \cdot FeCO_3$ – 10% mass,
sulfides and sulfates: pyrite FeS_2, gypsum $CaSO_4 \cdot 2H_2O$ – 25% mass,
oxides: silica SiO_2, hematite Fe_2O_3 – 15% mass.

During coal heating some of these components undergo decomposition or reactions with organic matter. The term ash usually means the substance that remains after combustion in standard temperature (815°C or 750°C). As result of such combustion about 10% of the mineral substance is lost to the gas phase. Gumz (1962) presented a simple formula enabling the calculation of mineral matter content in coal as a function of ash and total sulphur content

$$\text{mineral matter} = 1.11A + 0.35\, S_t \qquad (1.1)$$

Given et al. (1975) developed a formula in which the mineral matter content is related to pyritic sulphur and cloride content

$$\text{mineral matter} = 1.13A + 0.47\, S_{pyr} + 0.5\, Cl \qquad (1.2)$$

Some 97% of ash mass is formed by the following oxides: SiO_2, $Al_2O_3 + TiO_2$, Fe_2O_3, CaO, MgO, K_2O, Na_2O, SO_3, and P_2O_5. The remaining mass contains the so called trace elements: Zn, Pb, V, Ni, Cu, Co, and Cr.

The behaviour of mineral matter during combustion is characterized by the temperature of specific points during slow heating ($\sim 8K/min$) of fabricated standarized shape ash samples (cone, piramid or cube):

Sintering temperature — at which the first smelting of edges occurs associated with slight decrease of sample size without any deformation.

Initial deformation temperature — at which the first rounding of apex of the sample occurs.

Softening temperature — at which the sample has fused down to a spherical shape of which the height is equal the width of the base.

Hemispherical temperature $T_{1/2}$ — at which the height of the fused down sample is one half of the width of the base.

Fluid temperature — at which the fused sample is spread out in a nearly flat layer with a maximum height of 1/3 of the hemisphere.

Usually only four of the above temperatures are used, depending on the tradition of the country. In Anglo-american literature the sintering temperature is not used, while in many European countries the softening temperature is not used. Figure 1.3. presents the image of a standarized ash sample at three most often used characteristic temperatures for two shapes of the sample.

Generally coal ashes with high value of fluid temperature are benefical for combustion chambers with solid state ash removal, while those with low value of fluid temperature should be used in combustion chambers with liquid state ash removal. The difference between the characteristic temperatures is important, because it informs of temperature range in which ash has a tendency to agglomerate, stick to the walls, or flow down. Should this difference be small or large depends on the type of combustor.

The viscosity of liquid ash is an important parameter allowing us to determine the ability of ash to flow down the walls and out of the combustor bottom. A temperature of critical viscosity T_{cv} describing a point during cooling of liquid ash at which the viscosity begins to increase rapidly, is used. Sage and McIlroy (1960) present a simple formula

$$T_{cv} = T_{1/2} + 111°C \qquad (1.3)$$

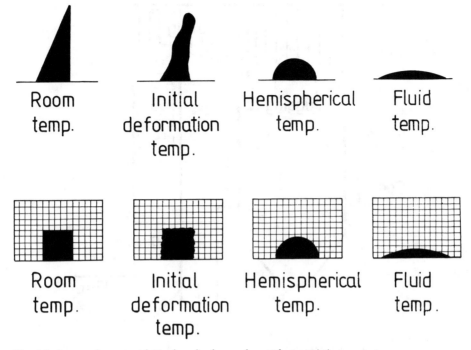

Fig. 1.3. Image of two standarized coal ash samples at characteristic temperatures.

Watt (1969), for a normalized ash composition

$$SiO_2 + Al_2O_3 + Fe_2O_3 + CaO + MgO = 100 \qquad (1.4)$$

proposed equation for temperature of critical viscosity in °C

$$T_{cv} = 2990 - 1470 \frac{SiO_2}{Al_2O_3} + 360 \frac{SiO_2}{Al_2O_3}$$
$$- 14.7 (Fe_2O_3 + CaO + MgO)$$
$$+ 0.15 (Fe_2O_3 + CaO + MgO) \qquad (1.5)$$

At temperatures below T_{cv} the value of viscosity depends on the furnace atmosphere, which is less important at temperatures above T_{cv}. Watt and Fereday (1969) for the normalized composition (1.4) developed an equation allowing them to calculate the ash viscosity in (Pa ∘ s) units at temperature T in °C

$$\lg \eta = \frac{10\,M}{T - 150} + C, \ \text{Pa} \circ \text{s} \qquad (1.6)$$

where

$$M = 0.00835\,SiO_2 + 0.00601\,Al_2O_3 - 0.109 \qquad (1.7)$$
$$C = 0.00415\,SiO_2 + 0.00192\,Al_2O_3 + 0.00276\,Fe_2O_3$$
$$+ 0.00160\,CaO - 0.392 \qquad (1.8)$$

where the ash components are in % of mass content.

Because the maximum value of viscosity, at which ash can easily be removed from the bottom of a combustor is equal 25 Pa ∘ s, then the temperature T_{25}

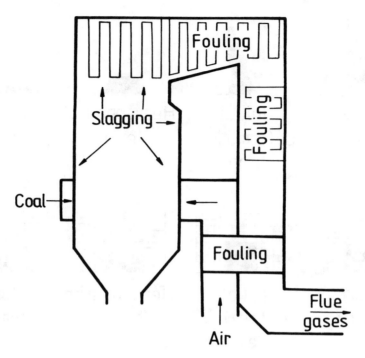

Fig. 1.4. Slagging and fouling surfaces in a combustor.

at which the viscosity of ash is equal 25 Pa ∘ s, is important. This temperature is then equal

$$T_{25} = \left(\frac{10^6 M}{\lg 25 - C}\right)^{1/2} + 150, \,^\circ C \qquad (1.9)$$

The behaviour of ash when contacting the cold surfaces of tubes within the combustion chamber or in convection ducts depends on the ash properties and on the gas dynamics of the combustor, both of which are equally important. The inorganic minerals, transformed into ash during combustion, may deposit onto heat transfer surfaces. This process is referred to as slagging if the deposit is in a molten and highly viscous state or fouling if the deposit is built by condensed species that vaporized earlier during combustion. Figure 1.4. presents the positions of slagging and fouling of heat transfer surfaces in a boiler.

The slagging tendency of coal ashes in which $Fe_2O_3 > CaO + MgO$ can be calculated by the following equation (Attig and Duzy, 1969)

$$R_s = \frac{Fe_2O_3 + CaO + MgO + K_2O + Na_2O}{SiO_2 + TiO_2 + Al_2O_3} S_{(wf)} \qquad (1.10)$$

where $S_{(wf)}$ means a mass content in % of sulphur in water free state coal. The value of the slagging index R_s is used as a basis for coal ash classification:

$$R_s > 0.6 \qquad \rightarrow \text{weak slagging tendency,}$$
$$R_s = 0.6 - 2.0 \rightarrow \text{moderate slagging tendency,}$$
$$R_s = 2.0 - 2.6 \rightarrow \text{high slagging tendency,}$$
$$R_s > 2.6 \qquad \rightarrow \text{severe slagging tendency.}$$

The fouling tendency of coal ashes containing $Fe_2O_3 > CaO + MgO$ is

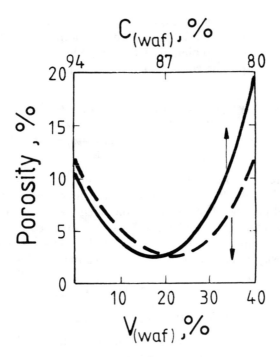

Fig. 1.5. Coal porosity.

characterized by a fouling index (Winegartner, 1974)

$$R_f = \frac{Fe_2O_3 + CaO + MgO + K_2O + Na_2O}{SiO_2 + TiO_2 + Al_2O_3} Na_2O \qquad (1.11)$$

where Na_2O must be taken in % of a normalized ash composition (eq. 1.4.).
For ashes with composition $Fe_2O_3 < CaO + MgO$ it is proposed to calculate
the fouling index R_f' by the same equation with the modification that instead
of Na_2O a water solveable Na_2O is taken (Winegartner, 1974). A following
classification of coal ashes is proposed:

R_f	R_f'	
< 0.2	< 0.1	→ weak fouling tendency,
0.2–0.5	0.1–0.25	→ moderate fouling tendency,
0.5–1.0	0.25–0.7	→ high fouling tendency,
>1.0	>0.7	→ severe fouling tendency.

1.3. COAL POROSITY

All coals have porous structure which is identified by volume, surface area,
and size distribution of pores. The access of gaseous agents to the inner particle surfaces depends on the size distribution of pores, classified most often
into three groups:

micropores	—	pores of diameters less than 2 nm,
mezopores	—	pores of diameters 2–50 nm,
macropores	—	pores of diameters bigger than 50 nm.

It is worth noting that above 90% of the particle internal surface area is
formed by micro/mezopores.

Figure 1.5. presents the variation of coal porosity as a function of carbon

and volatile matter content in "waf" state coal based on data by King and
Wilkins (1944). Surprisingly, the porosity does not increase with volatile content but the curves show minimum at $C_{(waf)} \approx 88\%$ or $V_{(waf)} \approx 20\%$.

The density of porous coal or carbonaceous particle can be determined
in two ways:

> true density — a ratio of the coal mass to the solid body volume
> (particle volume minus volume of pores),
>
> apparent density — a ratio of the coal mass to the volume of the par-
> ticle outside contour.

The total volume of pores is measured by helium (He) which because of its
atom size has the best chance to penetrate the whole porous structure (> 0.4
nm) of the coal particle. The volume of the outside contour of the particle is
measured by mercury (Hg) under 0.1 MPa pressure, because it cannot then
penetrate pores smaller than 15 μm. Coals and chars show good correlation
of the true density ρ (He) with the hydrogen content

$$\frac{1}{\rho(He)} = 0.4397 + 0.1223\ H_{(waf)} - 0.01715\ H_{(waf)}^2$$
$$+ 0.001077\ H_{(waf)}^3, \quad cm^3/g\ (waf) \qquad (1.12)$$

where hydrogen content must be taken in %. Equation (1.12) can be successfully applied within hydrogen content 0–7.7% (Johnson, 1978).

1.4. HEAT OF COMBUSTION

It has been stated that the effect of chemical bonds of the coal organic matter
on the heat of combustion is small. Consequently then it is possible to relate
the heat of combustion to the elementary analyses of coal. A number of
expressions are proposed which are derived statistically on the basis of the
measured values of heat of combustion. Masson and Gandhi (1983) tested
five most often quoted equations for the heat (upper) of combustion H_u for
coals and chars in dry state and standard temperature T_n:

a. Dulong

$$H_{u,n} = 33829\ C_{(wf)} + 144277\ H_{(wf)} + 9420\ S_{(wf)} - 18036\ O_{(wf)} \quad (1.13)$$

b. Grummel and Davies (1933)

$$H_{u,n} = \left(\frac{15219\ H_{(wf)}}{1 - O_{(wf)}} + 98767\right)\left(\frac{C_{(wf)}}{3} + H_{(wf)} - - \frac{O_{(wf)}}{8} + \frac{S_{(wf)}}{8}\right) \quad (1.14)$$

c. Mott and Spooner (1940)

$$O_{(wf)} \leq 15\%$$
$$H_{u,n} = 33620\ C_{(wf)} + 141933\ S_{(wf)} - 14528\ O_{(wf)} \qquad (1.15)$$
$$O_{(wf)} > 15\%$$
$$H_{u,n} = 33620\ C_{(wf)} + 141933\ H_{(wf)} + 9420\ S_{(wf)}$$
$$- \left(15324 - \frac{7201\ O_{(wf)}}{1 - A_{(wf)}}\right) O_{(wf)} \qquad (1.16)$$

d. Boie (1953)

$$H_{u,n} = 35169\ C_{(wf)} + 116247\ H_{(wf)} + 10467\ S_{(wf)}$$
$$- 11095\ O_{(wf)} + 6280\ N_{(wf)} \qquad (1.17)$$

e. Mason and Gandhi (1983)

$$H_{u,n} = 34095\, C_{(wf)} + 132298\, H_{(wf)} + 6848\, S_{(wf)} - 1531\, A_{(wf)}$$
$$- 11996\, (O_{(wf)} + N_{(wf)}) \tag{1.18}$$

where the heat of combustion is in kJ/kg(wf) and the coal composition in decimal fractions. During testing for variety of carbonaceous substances the standard error for formula (1.18) was about twice lower than for the others (1.13–1.17).

1.5. SPECIFIC HEAT

Van Krevelen (1981) proposed to calculate the specific heat of organic matter of coal at ambient temperature, by equation

$$c = \frac{R}{M} \tag{1.19}$$

where M is a molar weight of coal, calculated on basis of elementar analysis

$$\frac{1}{M} = \sum \frac{g_i}{M_i} \tag{1.20}$$

Merrick (1983), inspired by van Krevelen, described the specific heat of carbonaceous substance by equation

$$c = \frac{R}{M} f(T) \tag{1.21}$$

As a function $f(T)$ in equation (1.21) Merrick adapted the Einstein expression

$$f(T) = 3 \exp\left(\frac{\theta}{T}\right) \left(\frac{\theta}{T}\right)^2 \bigg/ \left(\exp\left(\frac{\theta}{T}\right) - 1\right)^2 \tag{1.22}$$

where the Debye characteristic temperature is calculated from the condition $f(T_{ambient}) = 1$, which is fulfilled for $\theta = 1200$ K. More accurate results can be obtained if two characteristic temperatures were taken: $\theta_1 = 380$ K and $\theta_2 = 1800$ K. That gives the specific heat

$$c = \frac{R}{M} \frac{1}{3} (f_1(T) + 2f_2(T)) \tag{1.23}$$

For a carbonaceous substance containing organic matter, mineral matter and moisture, Merrick proposed to calculate the specific heat as equal

$$c = \sum_i g_i c_i \tag{1.24}$$

with recommendation for ash the Kirov equation (1.26). Equation (1.23) results in large error (about 30%) of specific heat value, particularly in the region of most intensive devolatilization.

Kirov (1965) developed a relation between the specific heat of carbonaceous substances and the amount of volatiles and temperature

$$c = (1 - A_{(wf)})c_o + A_{(wf)}c_a \tag{1.25}$$

where the specific heat of mineral matter (ash) is equal

$$c_a = 594 + 0.586\, T \tag{1.26}$$

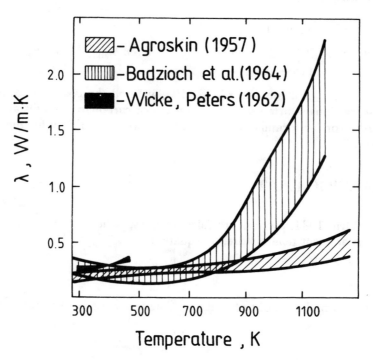

Fig. 1.6. Thermal conductivity of coal during heating.

while the specific heat of organic matter can be calculated as a function of the volatile matter in original coal:

for $V_{(waf)} > 0.1$

$$c_0 = (1 - V_{(waf)})c_1$$
$$+ (V_{(waf)} - 0.1)c_2 + 0.1 \, c_3 \qquad (1.27)$$
$$c_1 = -218 + 3.81 \, T - 1.76 \cdot 10^{-3} \, T^2 \qquad (1.28)$$
$$c_2 = 728 + 3.391 \, T \qquad (1.29)$$
$$c_3 = 2273 + 2.55 \, T \qquad (1.30)$$

and for $V_{(waf)} < 0.1$

$$c_o = (1 - V_{(waf)})c_1 + V_{(waf)} \, c_3 \qquad (1.31)$$

For a low rank coal ($V_{(waf)} = 32.8\%$, $C_{(waf)} = 77.0\%$, $H_{(waf)} = 5.8\%$, $O_{(waf)} = 14.9\%$) very good results can be obtained by an equation developed on the basis of Agroskin et al.(1970) data

$$c = 1.15 + 2.03 \cdot 10^{-3}(T - 300) - 1.55 \cdot 10^{-6}(T - 300)^2 \qquad (1.32)$$

Equations (1.25–1.32) give specific heat in J/(kg K) if temperature is used in K.

1.6. THERMAL CONDUCTIVITY

Thermal conductivity of carbonaceous porous substance is determined by conduction of heat through the solid body, conduction through the gas volume within the pores, and radiation of heat between the walls of pores. The contribution of radiation increases rapidly with temperature over 900 K, particularly in particles of large porosity. Figure 1.6. presents thermal conductivity measured by Agroskin (1957), Badzioch et al. (1964), and Wicke and Peters (1968). The Badzioch et al. data are very often quoted in literature, but they include some uncertainty resulting from the fact that their thermal conductivity together with the thermal diffusivity values lead to specific heat decreasing with temperature. For a subbituminous coal very good predictions of particle temperature can be obtained by the Agroskin thermal conductivity

$$\lambda = 0.19 + 2.5 \cdot 10^{-4}(T - 300) \qquad (1.33)$$

where with temperature in K we get λ in W/(m K).

1.7. THERMAL DECOMPOSITION OF COAL

Heating of coal causes its decomposition called in literature devolatilization or pyrolysis. It is important to notice that both the organic and the mineral parts of coal undergo decomposition. First the desorption of gases stored in pores of coal structure during its formation, takes place. The desorption of water steam, methane, carbon dioxide, and nitrogen begins at temperatures about 100°C. Above 300°C the release of liquid hydrocarbons, called tar, becomes important. Together with tar, gaseous hydrocarbons, carbon dioxide, carbon monoxide, and water steam, are released. At temperatures below 400°C the coal particle remains essentially unchanged, while above 400°C coals undergo a plastic state during which the shape and the size of the particle can change drastically. Above 550°C the plastic state ends and coal becomes again a hard substance, called char. During further heating hydrogen and carbon monoxide are released and char stabilizes its structure.

The chemical processes during coal decomposition are presented in literature by many mechanisms, among which the simplest is that proposed by Van Krevelen (1981):

$$\text{Coal} \rightarrow \text{Metaplast} \qquad (1.34)$$
$$\text{Metaplast} \rightarrow \text{Primary char} + \text{Primary volatiles} \qquad (1.35)$$
$$\text{Primary char} \rightarrow \text{Char} + \text{Secondary gas.} \qquad (1.36)$$

The whole mechanism contains three reactions. In temperature above 300°C coal depolymerises in a nonstable contemporary state, called "metaplast" by Van Krevelen. In the second reaction metaplast splits into primary char and primary volatiles (tar + primary gas). The nonstable primary char stabilizes in the third reaction releasing secondary gases.

Coal devolatilization is influenced by both the physical and chemical processes:

heat transfer to the coal particle and within it,
chemical primary reactions leading to products formation,
diffusion of products through the smelted or porous structure of coal,
chemical secondary reactions between products or between products and solid char.

The primary chemical reactions can be described by kinetic equations and are independent of pressure. Both the secondary chemical reactions and the diffusion processes are strongly dependent on pressure. The transport of products through the plastic body or through the pores is responsible for the formation of the final porous structure of char. During this transport the cracking of tar within the pores can take place leading to formation of lower molecular tar and gases as well as deposition of solid phase within the pores according to reactions:

$$Liquid \rightarrow Liquid + Gas, \tag{1.37}$$
$$Liquid \rightarrow Solid + Liquid \tag{1.38}$$
$$Liquid \rightarrow Solid + Gas. \tag{1.39}$$

The residence time of tar within the coal, depending on pressure, is the main factor influencing the extent of cracking reaction.

During the devolatilization the following homogeneous and heterogeneous reactions can also take place:

$$Methanation, \; CO + 3H_2 \quad \rightarrow CH_4 + H_2O, \tag{1.40}$$
$$Hydrogasification, \; C + 2H_2 \rightarrow CH_4, \tag{1.41}$$
$$Water \; gas, \; CO + H_2O \quad \rightarrow CO_2 + H_2 \tag{1.42}$$
$$Boudouard, \; C + CO_2 \quad \rightarrow 2CO \tag{1.43}$$

The first two reactions accelerate with increasing pressure, the third is neutral, while the fourth slows down. It is important to notice that the mineral matter in coal can influence all these reactions.

The viscosity of the plastic coal depends not only on temperature but also on duration time of devolatilization; consequently then its value can vary in a broad region from viscosity of sticky fluids, some $10^3 (Pa \circ s)$ to over $10^{14} (Pa \circ s)$. Van Krevelen (1981) proposed the plasticity of a carbonaceous substance to be proportional to the content M of the contemporary nonstable metaplast. The simplest kinetic equations describing the metaplast M and volatile V contents have a form:

$$\frac{dM}{dt} = k_V V - k_M M \tag{1.44}$$

$$\frac{dV}{dt} = -k_V V \tag{1.45}$$

Assuming $k_V \approx k_M$, we get a simple kinetic form

$$\frac{dM}{dt} = k_V(V - M) \tag{1.46}$$

The viscosity of the plastic coal is then inverse proportional to the metaplast content.

Decomposition of coal is not thermally neutral. During heating up the subsequent reactions have either exothermic or endothermic character. Thermal effects during coal heating are generated both in the organic matter as well as in the mineral matter. Figure 1.7. presents thermogravimetrical results, evaluated in form of enthalpy of devolatilization for subbituminous coal ($V_{(wf)} = 32.8\%$, $A_{(wf)} = 14.7\%$, $C_{(waf)} = 77.0\%$, $H_{(waf)} = 5.8\%$, $O_{(waf)} = 14.9\%$) obtained for two rates of heating. The exothermic effects can be observed already at temperatures above 200°C. Within 500–750°C region the process is endothermic, and again above 750°C exothermic. The total thermal effect is certainly very small if the final temperature of devolatilization is on the level of 700–800°C. Because during the combustion process particles are heated well above 800°C, then the total effect is exothermic. It has to be pointed out that the thermal effect of mineral matter decomposition is very strong. This is the main reason why data from various coals differ from each other. The total heat that has to be supplied in order to decompose coal to

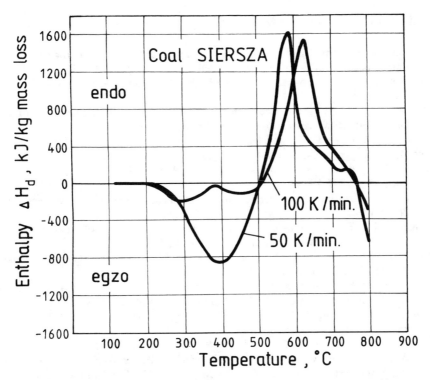

Fig. 1.7. Enthalpy of thermal decomposition of coal Siersza for two rates of heating
(Tomeczek and Palugniok, 1990).

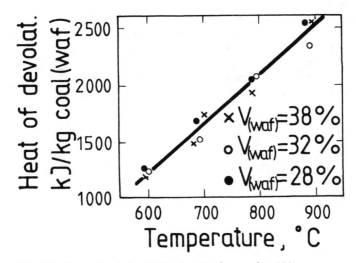

Fig. 1.8. Heat of coal devolatilization (Verfuss et al., 1982).

products having the final temperature of heating, can be calculated by equation

$$Q = \int_{T_n}^{T} mcdT + \int_{m_{initial}}^{m_{final}} \Delta H_d \, dm \qquad (1.47)$$

Verfuss et al. (1982) measured the heat of coal decomposition during fast
heating. Figure 1.8. presents the results to be almost independent of the volatile content within the region 28–38% (waf).

REFERENCES TO CHAPTER 1

Agroskin A. A.—The change of heat and temperature transfer coefficient of coal during heating (in German). Bergakademie Freiberg, J.9 (1957)177–186.

Agroskin A. A., Gonczarow E. I., Makeev L. A. and Jakunin W. P.—Koks i Chimija, 5 (1970) 8–13.

Attig R. C. and Duzy A. F.—Coal Ash Deposition Studies and Application to Boiler Design. American Power Conference, April 22, 1969.

Badzioch S., Gregory D. R., and Field M. A.—Fuel, 43 (1964) 267–280.

Boie W.—Energietechnik, 3 (1953) 309–316.

Given P. H., Cronauer D. C., Spuckman W., Lovel H. L., Davies A. and Biswas B.—Fuel, 54 (1975) 40–48.

Gumz W.—Kurzes Handbuch der Brennstoff und Feuerungstechnik. Springer-Verlag. Berlin 1962.

Johnson J. L.—Fundamentals of Coal Gasification in Chemistry of Coal Utilization. Chapter 23, John Wiley and Sons. New York, 1981.

King J. G. and Wilkins E. T.—In. Proc. Conf. Ultra-fine Structure of Coals and Cokes. B.C. U. R. A. London (1944)46–56.

Kirov N.Y.—Specific heats and total heat content of coals and related materials at elevated temperatures. B.C.U.R.A. Month. Bull., Vol. 29 (1965)33–57.

Mason D. M. and Gandhi K. N.—Fuel Processing Technology, 7(1983)11–22.

Merrick D.—Fuel, 62(1983)540–546.

Mott R. A. and Spooner C. E.—Fuel, 19(1940) 226–231, 242–251.

Sage W. L. and Mc Ilroy J. B.—Journ. Engineering for Power, April (1960), 145–155.

Stach E., Mackowsky M. Th., Teichmuller M., Taylor G. H., Chandra D. and Teichmuller R.—Coal Petrology. Gebruder Borntraeger, Berlin, Stuttgart, 1982.

Tomeczek J. and Palugniok H.—Proc. of Thermodynamics Conference. Krakow (1990)799–806.

Van Krevelen D. W.—Coal. Elsevier. Amsterdam, 1981.

Verfuss F., Lehman J. and Ahland E.—Erdöl und Kohle, 35(1982)332–336.

Watt J. D.—J. Inst. Fuel, 42(1969)131–134.

Watt J. D. and Fereday F.—J. Inst. Fuel, 42(1969)99–103.

Wicke M. and Peters W.—Brennstoffchemie, Bd.49(1968)97–102.

Winegartner E. C., Editor—Coal Fouling and Slagging Parameters. ASME, 1974.

2
COMBUSTION OF A
SINGLE COAL PARTICLE

Investigation of coal particle combustion in industrial conditions is not an easy task; therefore, most of the present knowledge was gained through experiments with single coal particles. In such cases the number of parameters influencing the process is limited to the following:

- type of coal,
- size of coal particle,
- rate of heating,
- gaseous atmosphere surrounding the particle.

It is worth mentioning that particles of coal can differ much in composition, even when prepared from one lump. Grinding of coal creates particles from almost pure organic matter of density 1.3 g/cm³ to almost pure mineral matter of density 2.6–5.0 g/cm³. In the case of pulverized coal most particles (up to 60%) contain only traces (<2%) of mineral matter. To obtain then valuable results it is necessary to reproduce several times the experiments with single particles. Another way is to use particles of one petrographic kind, which have more uniform composition.

2.1. TEMPERATURE FIELD WITHIN A COAL PARTICLE DURING HEATING

A coal particle dropped into a high temperature environment does not heat up uniformly. The rate of surface temperature increase and the profile of temperature within the particle are the function of:

- heat flux to the particle surface by radiation and convection,
- particle size,
- thermal properties of the particle,
- thermal effects within the particle.

To obtain a temperature profile within the particle it is necessary to solve the energy equation together with the boundary and initial conditions. For proximate analysis, a coal particle can be treated as an inert nonporous solid sphere, for which the solution is much easier. Field et al. (1967), based on the solution given by Carslaw and Jaeger (1959), present a simple equation for a maximum temperature difference within a particle of constant heat conduction coefficient λ_p heated from outside by a constant heat flux intensity $\dot{q}$

$$\Delta T_{max} = \frac{d_p \dot{q}}{8 \lambda_p} \qquad (2.1)$$

In a real coal particle, however, during heating the generated volatile matter can influence considerably the temperature distribution within the particle. Peters (1963) has shown that the flow of volatiles to the particle surface during intensive heating can reduce the convective heat transfer from sur-

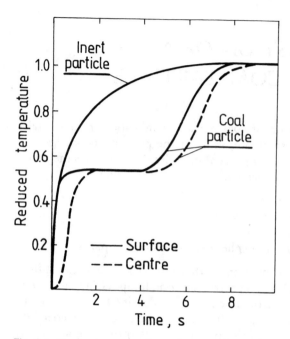

Fig. 2.1. Qualitative comparison of temperature within
heated coal particle (A) and inert particle (B).

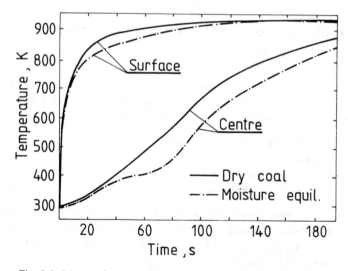

Fig. 2.2. Measured temperature runs in the centre and at the surface
of 15mm granules immersed in a 930K fluidized bed (Tomeczek and Kowol, 1991).

roundings to the particle surface. He suggested that the heat transfer coefficient to the particle surface decreases 10 times during fast heating of coal
particles mixed with hot solid heat carrier. Kallend and Nettleton (1966) on
the basis of experiments with 1–3 mm coal particles confirmed this. The
consequence of a decreased heat transfer to the particle surface is a temperature plateau of the particle surface on the level of about 400°C, which lasts

during the whole time of volatiles release. Davies and Brown (1969) also recorded the mentioned temperature plateau at the particle surface during the devolatilization time; they explained it, however, by a strong endothermic effect associated with devolatilization and not by a decreased heat transfer coefficient. A qualitative comparison of temperature variation of particle centre and of surface with characteristic temperature plateau recorded by Peters and Bertling (1965), and that calculated for inert nonporous particle is presented in figure 2.1.

Tomeczek and Kowol (1991) measured the temperature within 15 mm of coal granules dropped into a fluidized bed of 930°C temperature. Two kinds of granules were used: dry and containing moisture at equilibrium with ambient air. Figure 2.2 presents that at the centre of granules with equilibrium moisture there is a clear decrease of the rate of temperature increase at level of about 100°C due to evaporation of water, not observed for dry granules. No effects however, neither at the centre nor at the surface, are observed at the level of coal devolatilization (about 400°C). It means that neither the strong endothermic effects nor the decrease of heat transfer coefficient to the granule surface from the fluidized bed took place for the analyzed coal ($V_{(wf)}$ = 32.8%, $A_{(wf)}$ = 14.7%, $C_{(waf)}$ = 77.0%, $H_{(waf)}$ = 5.8%, $O_{(waf)}$ = 14.9%).

The accuracy of temperature field prediction within a coal particle depends on the knowledge of thermal properties of coal and their changes during devolatilization. Speaking of coal it is necessary to remember the changing nature of the substance. Already above 100°C, and certainly above 200°C, the term coal should not be used.

2.2. COAL PARTICLE SWELLING DURING HEATING

Heated coal particles undergo swelling. The scale of size change depends mainly on the:

- type of coal,
- initial particle diameter,
- composition of the surrounding gases,
- rate of heating.

The free swelling index, determined in standard conditions, does not indicate the size increase of a coal particle under conditions of rapid heating. It is worth noting that small particles (< 100 μm) behave differently than larger particles of about several milimeters. Generally, rapid heating leads to bigger swelling. Stahlherm et al. (1974) observed that high volatile coals produce particles of larger porosity in result of intensive release of gases. The particles of low volatile coals after devolatilization have small porosity and burn on the outer surface. It has been established that swelling of coal heated in inert gas is bigger than in air (Street et al., 1969). Figures 2.3 and 2.4 present the size variation of subbituminous coal particles heated in inert atmosphere. It can be seen that the bigger particle starts to expand when the surface temperature has reached the level of over 600°C, while the smaller is already expanding at surface temperature 300°C. This indicates the importance of temperature field within the particle on its behaviour. We can see also two different patterns of size variation. The smaller particle which was

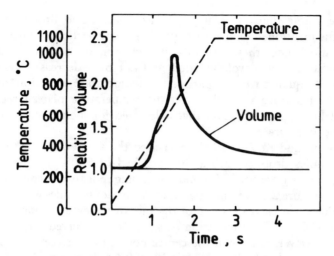

Fig. 2.3. Coal particle (0.4mm) volume variation during heating in argon (Tomeczek and Heród, 1979).

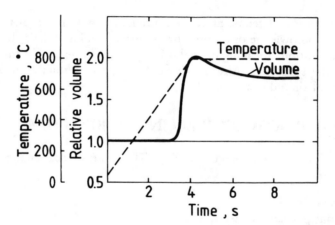

Fig. 2.4. Coal particle (0.81mm) volume variation during heating in argon (Tomeczek and Heród, 1979).

heated faster and to higher temperature did collapse almost to the initial size; the bigger particle which was heated more slowly and to lower temperature shrank only slightly after reaching the maximum size.

Because of the nonhomogeneous composition of coal particles, the results of experiments with single particles are mostly scattered. The experiments with macerals or microlithotypes give usually more uniform results. Figures 2.5 and 2.6 present the final expansion of particles of various sizes heated with the rate of 100 K/s to final temperature 1000°C in oxidizing atmosphere. The vitrite shows considerable expansion while the inertinite particles almost do not expand at all. This example shows clearly why it is so difficult to draw general conclusions from experiments with single coal particles.

Particles during swelling not only expand but also change their shape. A common behaviour is that nonspherical shapes with sharp edges become

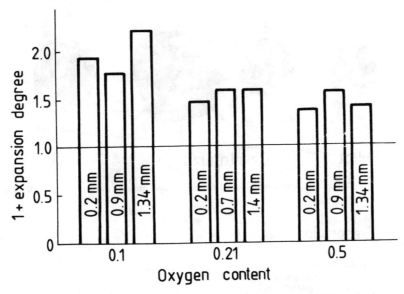

Fig. 2.5. Expansion of vitrite particles heated with the rate 100 K/s to final temperature 1000°C in three environments (Tomeczek and Heród, 1979).

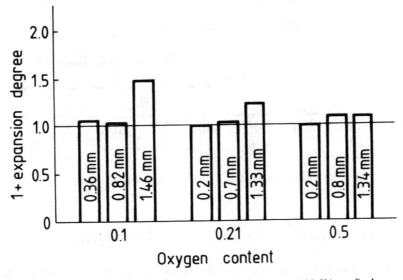

Fig. 2.6. Expansion of inertite particle heated with the rate 100 K/s to final temperature 1000°C in three environments (Tomeczek and Heród, 1979).

rounded off. Figure 2.7 presents the stages of particle shape during heating. Street et al. (1969) found pores on the outer surface of particles devolatilized in air more often than in nitrogen. They found also many particles without any pores on the outer surface.

Kaiser (1981), Arendt (1980), and Lowenthal (1984) recorded shifting of the position of maximum particle expansion towards higher temperature with increasing pressure. Also a diminishing of particle shrinkage in the final devolatilization phase is observed, so that at 9 MPa no shrinkage took place.

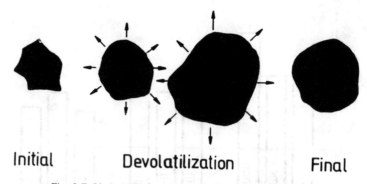

Fig. 2.7. Variation of coal particle shape during heating.

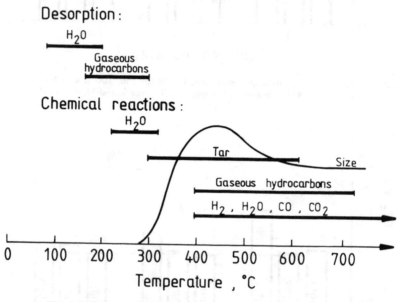

Fig. 2.8. Regions of volatiles release during coal particle heating.

2.3. KINETICS OF COAL DEVOLATILIZATION

The amount and composition of volatiles released during coal heating depend mainly on the type of coal and the final heating temperature. Devolatilization of coal in oxidizing atmosphere is associated with combustion of the released volatiles in the form of a diffusion flame surrounding the particle. The volatiles are composed of liquids and gases, which is somewhat confusing because during devolatilization both components are in gaseous phase when leaving the particle surface. The so called liquids contain hydrocarbons (tar) and water. It has been commonly stated that release of tar is completed within the temperature region 350–500°C, that is associated with the plastic state of coal and its swelling. Release of gases lasts to much higher temperature. Figure 2.8 presents the regions of volatiles release together with the curve of particle volume change during moderate rate of heating. Arendt et al. (1981) have shown that increased rate of heating causes shifting of the beginning of tar release towards higher temperature.

Up until now it is not possible to build a model of coal devolatilization on the basis of fundamental coal structure. The developed models with empirical coefficients do allow us, however, to determine the products of devolatilization but do not explain the nature of the process. These models postulate a set of reactions between the unspecified hypothetical components describing the formation of products. Although these models do not utilize the chemical structures and elementary reactions, they offer useful tools in the form of kinetic equations.

The simplest models with a single first or higher order reaction do allow us to obtain sufficient accuracy in cases when the curves of products formation have simple shape with single maximum. This demand can satisfactorily be fulfilled at high rates of heating, over 100 K/s, and in narrow temperature region. The rate of volatile products formation in a single reaction of N-th order from 1 kg coal (waf) can be described by equation

$$\frac{dm_V}{dt} = k_o \exp(-E/RT)(m_\infty - m_V)^N \qquad (2.2)$$

In cases when the rate of formation of individual j-th volatile product is demanded, it can be calculated by equation

$$\frac{dm_j}{dt} = k_{oj} \exp(-E_j/RT)(m_{\infty j} - m_j)^{N_j} \qquad (2.3)$$

The value m_∞ meaning the mass of products released after long residence ($t \to \infty$) of 1 kg of coal (waf) in temperature T, should not be identified by the volatiles content $V_{(waf)}$ measured in standard conditions. This value depends on the final temperature of heating. However, because a single reaction model assumes m_∞ to be a constant value it is a source of error when applying this model in broad temperature region.

The order of reaction N is equal (Van Heek, 1981):

$N = 0$—process is controlled by transport phenomena,
$N = 1$—reaction takes place uniformly in the whole
                volume of the solid,
$N = 2$—reaction is determined by bimolecular processes.

A one step mechanism with one chemical reaction would be the most convenient for mathematical modelling of combustion; however one must remember that the accuracy of such models is not high. In this group of models most frequently the data by Badzioch and Hawksley (1970) are quoted for a model

$$\text{Coal} \to \text{Solid Residue} + \text{Volatiles} \qquad (2.4)$$

in which the total amount of volatiles can be described by kinetic equation

$$\frac{dm_V}{dt} = k_{oV} \exp(-E_V/RT)(m_{\infty V} - m_V) \qquad (2.5)$$

where $m_{\infty V}$ can be assumed to be equal to the volatile content of coal and the preexponential factor and the activation energy are equal:

	highly swelling coals	weakly swelling coals
$k_{oV} =$	$6 \circ 10^5 \text{ s}^{-1}$	$1.5 \circ 10^5 \text{ s}^{-1}$
$E_V =$	74 kJ/mol	74 kJ/mol

Bliek et al. (1985) present a multi step model

$$\text{Coal} \begin{array}{c} \overset{1}{\nearrow} T \overset{4}{\rightarrow} \alpha R + (1 - \alpha)V_{1,4} \\ \overset{2}{\rightarrow} V_2 \\ \overset{3}{\searrow} V_3 \end{array} \qquad (2.6)$$

for which the kinetic equations have the form:

$$\frac{dm_T}{dt} = k_{o1}\exp(-E_1/RT)(m_{\infty 1} - m_T)^2 \qquad (2.7)$$

$$\frac{dm_{V1,4}}{dt} = k_{o1}\exp(-E_1/RT)(m_{\infty 1} - m_T) - k_{o4}$$
$$\circ \exp(-E_4/RT)m_T(1 - \alpha) \qquad (2.8)$$

$$\frac{dm_{V2}}{dt} = k_{o2}\exp(-E_2/RT)(m_{\infty 2} - m_{V2})^2 \qquad (2.9)$$

$$\frac{dm_{V3}}{dt} = k_{o3}\exp(-E_3/RT)(m_{\infty 3} - m_V)^2 \qquad (2.10)$$

For coal ($A_{(wf)}$ = 6.2%, $V_{(waf)}$ = 35.7%, $C_{(waf)}$ = 64.3%) the value α = 0.8 and the kinetic constants are equal:

$k_{o1} = 1.65 \cdot 10^{12}$ s^{-1}, E_1 = 188 kJ/mol, $m_{\infty 1}$ = 16.9% (waf),
$k_{o2} = 8.36 \cdot 10^3$ s^{-1}, E_2 = 72.2 kJ/mol, $m_{\infty 2}$ = 9.0% (waf),
$k_{o3} = 3.68 \cdot 10^6$ s^{-1}, E_3 = 142.5 kJ/mol, $m_{\infty 3}$ = 6.5% (waf),
$k_{o4} = 2.70 \cdot 10^4$ s^{-1}, E_4 = 60 kJ/mol.

Niksa et al. (1984) presented a two step mechanism

$$\text{Coal} \overset{k_1}{\rightarrow} \alpha V_1 + (1 - \alpha)I \underset{k_3}{\overset{k_2}{}} \begin{array}{c} \nearrow \text{Residue} \\ \searrow V_3 \end{array} \qquad (2.11)$$

Models (2.4) and (2.11) determine the mass of the volatiles but do not identify their chemical composition. That makes them difficult to use in combustion modelling. The multireaction kinetic models can be applied in broader ranges of conditions and describe the release of individual chemical components. The rate of j-th product formation in L independent reactions is given by equation

$$\frac{dm_j}{dt} = \sum_{l=1}^{L} k_{ojl}\exp(-E_j/RT)(m_{\infty jl} - m_{jl})^{N_{jl}} \qquad (2.12)$$

in which usually $k_{ojl} = k_{oj}$ and $N_{jl} = N_j$.

Table 2.1. presents kinetic constants developed by Suuberg et al. (1979), Pottgiesser (1980), and by Tomeczek and Kowol (1991).

Wójcik (1991) has simplified the kinetic data by Tomeczek and Kowol (1991) assuming that the volatiles contain only: CH_4, C_2H_6, CO, CO_2, H_2O, and C_6H_6. The mass of liquid hydrocarbons is represented by benzene (C_6H_6). The kinetic data of the simplified model are given in table 2.2.

2.4. HEAT AND MASS TRANSFER WITHIN A DEVOLATILIZING COAL PARTICLE

Devolatilization of coal starts at the outer surface of the particle. Volatiles generated within a thin layer close to the surface have a short distance to

Table 2.1 **Comparison of Kinetic Data of Coal Devolatilization.**

Product	Authors	$m_{\infty jl}$ %, $kg_{jl}/kg(waf)$	E_{jl} kJ/mol	k_o s^{-1}	N_{jl}
H_2	Pottgiesser (1980)	1.19 1.33	119 183	$2.92 \cdot 10^7$ $7.90 \cdot 10^9$	3
	Suuberg et al. (1979)	1.0	377	$1 \cdot 10^{17}$	1
	Tomeczek and Kowol (1991)	0.02 0.08 0.54 0.03	153 181 217 222	$5 \cdot 10^8$	1
CH_4	Pottgiesser (1980)	1.40 1.57	204 158	$1.86 \cdot 10^{13}$ $6.00 \cdot 10^8$	2
	Suuberg et al. (1979)	0.7 1.8	230 272	$1 \cdot 10^{13}$ $1 \cdot 10^{12}$	1
	Tomeczek and Kowol (1991)	0.01 0.07 0.47 0.89 0.07 0.50	91 121 131 151 162 176	$1 \cdot 10^7$	1
CO	Pottgiesser (1980)	0.12 0.51 0.85	156 166 239	$1.08 \cdot 10^{11}$ $1.61 \cdot 10^9$ $4.60 \cdot 10^{11}$	2
	Suuberg et al. (1979)	0.40 0.21	230 272	$1 \cdot 10^{13}$	1
	Tomeczek and Kowol (1991)	0.114 0.396 0.683 1.308 1.652 1.518	163 165 183 207 238 257	$1 \cdot 10^{10}$	1
CO_2	Pottgiesser (1980)	0.09 0.24 0.32 0.50	201 104 202 211	$3.5 \cdot 10^{11}$ $1.1 \cdot 10^4$ $1.3 \cdot 10^9$ $1.6 \cdot 10^8$	1
	Suuberg et al. (1979)	0.4 0.9	167 272	$1 \cdot 10^{13}$ $1 \cdot 10^{13}$	1
	Tomeczek and Kowol (1991)	0.361 0.120 0.682 1.248 0.878 0.486	101 115 130 150 168 188	$1 \cdot 10^7$	1
C_2H_6	Tomeczek and Kowol (1991)	0.092 0.119 0.026 0.020	150 162 183 214	$5 \cdot 10^8$	1
C_2H_4	Tomeczek and Kowol (1991)	0.071 0.055 0.015 0.040	150 164 190 217	$5 \cdot 10^8$	1
liquids	Tomeczek and Kowol (1991)	4.2 8.2 3.8 3.6	155 201 225 312	$1 \cdot 10^{12}$	1

Table 2.2 **Kinetic Data of Coal Devolatilization.**

Product	k_o s^{-1}	E kJ/mol	m_∞ %,kg/kg(waf)
CH_4	$1.5 \cdot 10^4$	85.0	2.01
C_2H_6	$1.5 \cdot 10^3$	65.0	1.11
C_6H_6	$1.5 \cdot 10^7$ $1.0 \cdot 10^{12}$	120.0 312.0	12.00 3.60
CO	$5.0 \cdot 10^3$	75.0	5.67
CO_2	$1.75 \cdot 10^2$	40.0	4.85
H_2O	$1.0 \cdot 10^{12}$	312.0	4.20

flow in order to escape to the surroundings. However, as the devolatilizing layer proceeds towards the centre of the particle, the distance to cover increases. The flow of volatiles through the porous structure of devolatilizing coal and the layer of char formed from the outside surface of the particle is very complicated and difficult to describe. Most of the volatiles are generated on the walls of micropores and are then transported through a sequence of larger pores to the outside surface of the particle. According to Gavalas (1982) the lack of reliable physical properties of molten coal constitutes a formidable problem, particularly for plastic coal in which the bubbles of volatiles diffuse and collide through a molten coal with original pores smelted. Considerations presented in this chapter are limited therefore to nonplastic coals which preserve the original pore structure.

Opinions about the influence of the intraparticle mass transport on the devolatilization are divided. Howard and Essenhigh (1967) acknowledged the flow of volatiles from the inside of the particle through the porous char layer as the main factor influencing the devolatilization during rapid heating of the particle. Gavalas and Wilks (1980) and Simons (1984) assumed that both mass transport as well as chemical processes determine the rate of devolatilization. Koch et al. (1969) and Schwandtner (1971), however, assumed that the formed char layer creates practically no resistance to the flowing volatiles. A comprehensive model of heat and mass transfer during single particle devolatilization has been published by Bliek et al. (1985), which however does not allow for prediction of the evolution rate of individual gaseous species.

In order to describe the flow of volatiles through the porous structure of coal the following assumptions are made:

- for coal particles larger than 1 mm almost all the volatiles flow through pores of 1 μm diameter (average of 0.3–3 μm) into a group of 6 μm pores (average of 3–10 μm) that conduct the volatiles to the outer surface (Gavalas and Wilks, 1980),
- the pore structure does not change during devolatilization for nonplastic coal,
- liquid products have properties of benzene,
- flow of volatiles is convective quasi-stationary,
- chemical reactions in macropores 1 μm and 6 μm have a minor effect on the process.

The energy equation has a form

$$\frac{1}{r^2}\frac{\partial}{\partial r}\left(\lambda_p r^2 \frac{\partial T}{\partial r}\right) + \theta_6 \, w_{6,v} \, \rho_{6,v} \, c_{p,v} \frac{\partial T}{\partial r} + \frac{\partial Q_{ch}}{\partial t} = \rho_p \, c_p \frac{\partial T}{\partial t} \qquad (2.13)$$

The heat of pyrolysis in the energy equation (2.13) is assumed to be endothermic and proportional to the mass of the generated liquid products

$$\frac{dQ_{ch}}{dt} = -\Delta H_d \, \rho_{po(waf)} \frac{\partial m_1}{\partial t} \qquad (2.14)$$

This is justified by the fact that the thermal endothermic reactions are almost completed within the region of liquid products formation.

The mass balance of volatile products flowing through 6 μm pores is described by equation

$$\theta_6 \frac{1}{r^2} \frac{\partial (r^2 w_{6,v} \rho_{6,v})}{\partial r} = \rho_{po(waf)} \sum_{i=1}^{7} \frac{\partial m_i}{\partial t} \qquad (2.15)$$

The generated mass flux of each of the six gaseous (H_2, CH_4, CO, CO_2, C_2H_4, C_2H_6) and of the liquid volatile products is described by either four or six independent first order reactions

$$\frac{\partial m_i}{\partial t} = \sum_{l=1}^{4 \text{ or } 6} k_{oi} \exp(-E_{il}/RT)(m_{\infty il} - m_{il}) \qquad (2.16)$$

with kinetic parameters by Tomeczek and Kowol (1991) given in table 2.1.

The pressure distribution within the pores of diameter $d_6 = 6$ μm can be obtained by solution of the energy equation for volatiles flowing through the 6μm pores

$$-dp_6 = (w_{6,v} \rho_{6,v}) dw_{6,v} + \frac{32}{d_6^2} (w_{6,v} \rho_{6,v}) \nu_{6,v} \tau_6 \, dr \qquad (2.17)$$

The initial conditions for equations (2.12)–(2.17) are:

$$T(r,0) = T_n, \; p_6(r,0) = p_1(r,0) = p_n, \; g_i(r,0) = 0 \qquad (2.18)$$

The boundary conditions are:

- the gradients of temperature, velocity and pressure in the centre are equal to zero,
- heat flux from the surroundings to the outer surface is the sum of radiative and convective flux.

The pressure distribution within the pores of 1 μm diameter was obtained by equation (2.17) modified to have a subscript 1. The flow of volatiles in the 1μm pores is assumed to be isothermal so the kinematic viscosity does not change along the 1μm pore length.

Solution of equations (2.13)–(2.17) was obtained by a numerical procedure. Figure 2.9 presents comparison of the measured temperature runs in the centre and at the surface of a 5.3 mm dry coal granule dropped into a fluidized bed of 900°C temperature with the values calculated by the model in which the thermal and structural properties were equal: thermal conductivity—eq. (1.33), specific heat—eq. (1.32), 1 μm pore volume share $\theta_1 = 0.015$, 6 μm pore volume share—$\theta_6 = 0.019$, average length of 1 μm pores—

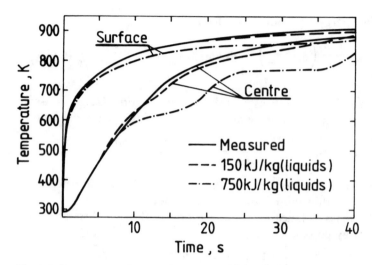

Fig. 2.9. Temperature runs in the centre and at the surface of dry coal granules of diameter 5.3mm for various endothermic enthalpy of devolatilization (Tomeczek and Kowol, 1991).

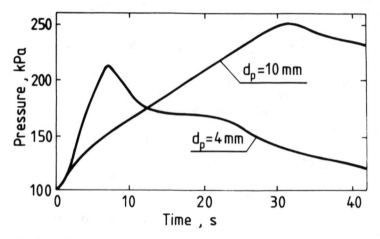

Fig. 2.10. Variation of pressure within 1μm pores at coal granules' centre during devolatilization in 1200K fluidized bed (Tomeczek and Kowol, 1991).

0.158 mm, tortuosity of 6μm pores—τ_6 = 1.5, coal density—ρ_p = 1170 kg/m^3, coal porosity—0.249, char porosity—(0.249–0.45). For the strong endothermic enthalpy of devolatilization 750 kJ/kg (liquids) a strong plateau in the granule's centre can be observed. The shape of the calculated temperature curve closest to the experimental curve have the results calculated for an endothermic enthalpy of 150 kJ/kg (liquids). For other granule diameters the most suitable value of endothermic enthalpy ΔH_d was in the range <200 kJ/kg (liquids).

Figure 2.10 presents the time functions of pressure in pores of 1μm diameter located at the centre of two granules calculated by the model. The highest pressure, about 0.25 MPa, occurs in the centre of a 10 mm granule at the time of intensive devolatilization. These values of pressure are not high

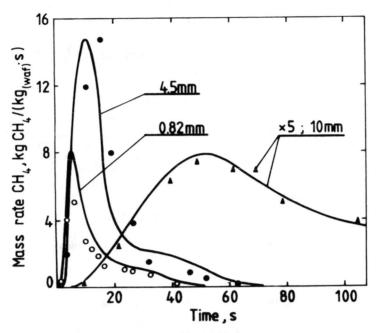

Fig. 2.11. Calculated mass flux of CH₄ released (solid line) from coal samples
dropped into hot fluidized beds compared with experimental results (Tomeczek and Kowol, 1991):
o—particles 0.75–0.9 mm, bed 873K;
●—particles 4–5 mm, bed 1150K;
▲—globules 10 mm, bed 870K.

enough to influence the kinetics of products formation and, as a consequence,
the temperature field within the particle.

The model can describe very well the release of gaseous products. Figure
2.11 demonstrates the accuracy of the model for three diameters of particles
heated in a fluidized bed. For the smallest particles the deviation between the
measured and calculated values at the point of maximum release is the largest;
this is due to the nonspherical shape of these particles.

The influence of intraparticle heat and mass transport on the devolatilization depends on the rate of heating and on the particle diameter. Even for
small particles ($100\mu m$) the temperature difference between the surface and
the centre can exceed several Kelvin at high rate of surface heating. Assuming
a linear variation of the surface temperature, it is possible to calculate by the
model the regions of kinetics-diffusion control of the devolatilization process.
Results are presented in figure 2.12 together with the curves by Koch et al.
(1969) and Schwandtner (1971). It can be seen that for high rates of heating,
over 10^3 K/s, the diffusion processes become important even for pulverized
coal.

The role of heat and mass transfer on the devolatilization process can
be also identified on the basis of relation between the time of devolatilization
and the particle diameter

$$t_V = B\, d_p^N \qquad (2.19)$$

in which for processes controlled by heat and mass transfer $N = 2$. For flui-

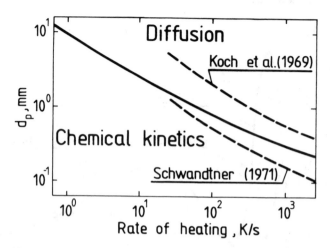

Fig. 2.12. Regions of kinetics-diffusion control of devolatilization (Kowol, 1985).

dized bed devolatilization during combustion a value of $N \approx 2$ is reported almost independently from coal type (Morris and Keairns (1970), Jung (1980)). Also for nonfluidized bed combustion a value of N close to 2 is reported:

Essenhigh (1963)—$N = 2.2$, $d_p = 0.3–4.1$ mm,
Ivanova and Babij (1966)—$N = 2$, $B = 4.5 \cdot 10^5$ s/m², $d_p = 0.1–1$ mm,
Ragland and Weiss (1979)—$N = 1.2$, $d_p = 2–12$ mm,
Kalend and Nettleton (1966)—$N = 2$, $B = 3.5 \cdot 10^6$ s/m², $d_p \leqslant 3$ mm.

Devolatilization time determined optically by the presence of volatiles flame does not have to be equal to the total devolatilization time calculated by means of the kinetic equations obtained on the basis of the mass loss measurement. This is due to the fact that the devolatilization process starts before the ignition of volatiles and continues after the disappearance of the volatiles flame.

2.5. MECHANISM OF SURFACE REACTIONS DURING COAL COMBUSTION

The mechanism of elementary reactions does allow for deeper understanding of the surface combustion process. The basis of this mechanism gives the theory of active centres, which postulates that reactions take place in favourable points of the surface, characterized by irregularities. The active centres can be related to: corner carbon atoms, inorganic inclusions or oxygen and hydrogen groups in the coal molecular structure. In each active centre the following elementar steps can take place:

- adsorption of gaseous agent,
- migration of the intermediate compound,
- desorption.

The rate of elementary surface reaction is proportional to the surface concentration of active centres $[C_t]$. The value of $[C_t]$ determines the reactivity of carbonaceous substance. The total concentration of active centres is a sum of surface concentrations of free active centres $[C_f]$ and of occupied active centres $[C(O)]$

$$[C_t] = [C_f] + [C(O)] \qquad (2.20)$$

The following ten reactions are most often considered in analysis of elementary surface reactions of carbonaceous substance combustion (Laurendeau, 1973):

— Reversible oxygen exchange reactions in which molecules of CO_2 or H_2O dissociate on free active centres

$$CO_2 + C_f \underset{k_{1b}}{\overset{k_{1f}}{\rightleftarrows}} CO + C(O) \qquad (R.1)$$

$$H_2O + C_f \underset{k_{2b}}{\overset{k_{2f}}{\rightleftarrows}} H_2 + C(O) \qquad (R.2)$$

— Desorption of carbon monoxide through dissociation of the $C(O)$ complex on the surface. This step breaks the structure of solids and leads to conversion of the inactive coal atom C_i into an active centre

$$C(O) + C_i \overset{k_3}{\rightarrow} CO + C_f \qquad (R.3)$$

— Dissociation of oxygen on free active centre and a following adsorption of oxygen atom

$$O_2 + C_f \overset{k_4}{\rightarrow} C(O) + O \qquad (R.4)$$

— Adsorption of oxygen atom

$$O + C_f \overset{k_5}{\rightarrow} C(O) \qquad (R.5)$$

— Transformation of active centre C_f into inactive centre C_i in high temperature annealing

$$C_f \overset{k_6}{\rightarrow} C_i \qquad (R.6)$$

— Dual adsorption of oxygen on active centres

$$O_2 + 2C_f \overset{k_7}{\rightarrow} 2C'(O) \qquad (R.7)$$

— Surface migration

$$C'(O) \overset{k_8}{\rightarrow} C(O) \qquad (R.8)$$

— Surface migration leading to conversion of inactive coal atom in an active centre

$$C(O) + C_i \overset{k_9}{\rightarrow} C(O) + C_f \qquad (R.9)$$

— Dual site desorption

$$C(O) + C(O) \overset{k_{10}}{\rightarrow} CO_2 + C_f \qquad (R.10)$$

The presence of water vapour and carbon dioxide influences the reaction rate, but the reaction with oxygen is so fast that this influence in surface reaction is negligible. Spokes and Benson (1967) proposed for typical combustion temperatures a mechanism based on reactions (R.3), (R.4) and (R.5). Assuming a constant surface concentration of $C(O)$ they obtained an equation describing the rate of surface reaction of char with oxygen

$$[\dot{R}] = m_C \frac{2[C_t]k_4 C_{O_2}}{1 + 2\dfrac{k_4}{k_3} C_{O_2}} \qquad (2.21)$$

Alternative mechanism was proposed by Simons and Lewis (1977)

$$O_2 + C_f \rightleftarrows C(O_2) \tag{2.22}$$

$$C(O_2) + C_f \rightarrow 2CO \tag{2.23}$$

The last equation could be responsible for the rate of surface combustion but evidence favours rather presence of $C(O)$ than $C(O_2)$.

Laurendeau (1979) suggests a mechanism based on reactions (R.3), (R.7), (R.8), (R.9), and (R.10). He has demonstrated that this mechanism does allow us to confirm the experimentally found values of reaction order $(0, \frac{1}{2}$ or $1)$ for surface reaction between char and oxygen dependent on the temperature level. At certain justified simplifications Laurendeau obtained the surface reaction rates at various temperatures:

low temperatures < 900 K, $k_3/k_8 \gg 1$, $k_9/k_8 \gg 1$

$$[\dot{R}] = m_C \frac{k_9[C_t]}{2} \tag{2.24}$$

that means reaction order $N = 0$ and the mobile centre desorption (R.9) controls the combustion,

intermediate temperature, 900 K $\leq T \leq 1500$ K

$$[\dot{R}] = m_C[C_t]k_8 \left(\frac{k_7}{k_{10}}\right)^{1/2} C_{O_2}^{1/2} \tag{2.25}$$

that means reaction order $N = 1/2$ and the centre migration (R.8) is the main controlling reaction,

high temperatures > 1500 K, $k_3/k_8 \gg 1$, $k_9/k_8 \ll 1$

$$[\dot{R}] = m_C 2[C_t]^2 k_7 C_{O_2} \tag{2.26}$$

that means the reaction order $N = 1$ and the dissociative hemisorption (R.7) controls the combustion.

2.6. EFFECTIVE RATE OF COAL PARTICLE COMBUSTION REACTION

Combustion process of a devolatilized coal particle is described by the following phenomena:

- diffusion of oxygen through the boundary layer surrounding the particle to the surface of the particle,
- diffusion of oxygen through the gases within the pores of the particle to the internal particle surface,
- chemical reactions on the outer and inner surfaces of the particle,
- diffusion of reaction products through the pores to the outer surface,
- diffusion of reaction products through the boundary layer to the bulk of the gases surrounding the particle,
- reaction of product gases within the boundary layer with oxygen diffusing to the particle.

Discussion in this chapter will be limited only to nonporous particles. Let us consider a simplified single stage mechanism of combustion

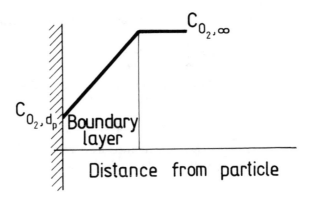

Fig. 2.13. Oxygen concentration profile in particle neighbourhood.

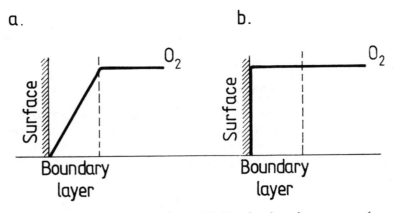

Fig. 2.14. Oxygen concentration in particle boundary layer for two cases of combustion control: a, by diffusion; b, by chemical reaction kinetics.

$C + O_2 \rightarrow CO_2$, presented in Figure 2.13. The rate of surface chemical reaction, assuming here to be of the first order, can be described by equation

$$\dot{R}_C = k_C A_p C_{O_2, d_p} \qquad (2.27)$$

The rate of oxygen diffusion through the boundary layer to the particle surface is equal

$$\dot{R}_D = \beta A_p (C_{O_2, \infty} - C_{O_2, d_p}) \qquad (2.28)$$

In cases when the rate of chemical reactions is much higher than the rate of oxygen diffusion $\dot{R}_C \gg \dot{R}_D$, that takes place at high solid surface temperature the oxygen concentration at the particle surface is equal zero $C_{O_2, d_p} = 0$. In opposite, for low solid surface temperature, the chemical reaction is much slower $\dot{R}_C \ll \dot{R}_D$ and the oxygen concentration at the particle surface is equal $C_{O_2, d_p} = C_{O_2, \infty}$. Figure 2.14 presents these two border situations. Most frequently we deal, however, with the intermediate case, for which we can write

$$\dot{R} = k A_p C_{O_2, \infty} \qquad (2.29)$$

The effective rate constant k of the reaction can be expressed by the rate constant of the chemical reaction and by the mass transfer coefficient

$$\frac{1}{k} = \frac{1}{k_C} + \frac{1}{\beta} \qquad (2.30)$$

The rate constant of surface chemical reaction is described by Arrhenius equation

$$k_C = k_o \exp(-E/RT_p) \qquad (2.31)$$

while the coefficient of mass transfer, defined as a ratio of the oxygen diffusion coefficient D and the boundary layer thickness δ

$$\beta = \frac{D}{\delta} \qquad (2.32)$$

can be calculated from the theory of the boundary layer by means of the Sherwood number

$$Sh = \frac{\beta d_p}{D} \qquad (2.33)$$

For a spherical particle the Sherwood number is equal

$$Sh = 2 + A \, Re^{0.5} Sc^{0.33} \qquad (2.34)$$

where $A = 0.552$ according to Frossling (1938) and $A = 0.68 (20 < Re < 2000)$ according to Rowe et al. (1965). The Reynolds number in the last equation should be calculated based on a relative gas particle velocity. For small particles $Sh \rightarrow 2$ and the rate of oxygen diffusion to the surface of the particle is equal

$$\dot{R}_D = 2D A_p \frac{C_{O_2,\infty}}{d_p} \qquad (2.35)$$

According to kinetic gas theory the diffusion coefficient depends on temperature according to function $T^{3/2}$. Many authors quote the power 1.78 (Andrussov, 1950) or 1.75 (Field et al., 1967). If the diffusion coefficient in standard conditions (p_n, T_n) is equal D_n, then in parameters (p, T) we have

$$D = D_n \left(\frac{T}{T_n} \right)^{1.75} \left(\frac{p_n}{p} \right) \qquad (2.36)$$

Because most frequently the solid particle temperature differs from the gas temperature, then the medium temperature of the boundary layer should be used in equation (2.36).

The effective rate constant of combustion reaction is a function of temperature which can be determined on the basis of equations (2.30), (2.31), (2.33), and (2.36). This function in coordinates $(\lg k, 1/T)$ is presented in figure 2.15. For low temperatures the process is controlled by chemical reaction kinetics. The slope of the curve depends on the activation energy, and the higher is the activation energy the larger is the slope. In diffusion controlled region the temperature dependence of the effective rate constant is smaller, as a consequence of relation (2.36). In reality there is no sharp change of slopes in the border point; there is always an intermediate region that causes a continuous change of gradients between the regions. It can be seen that the transfer from chemical reaction kinetics control to diffusion control takes place at various temperatures, because the mass transfer coefficient depends on the particle diameter $\beta = 2D/d_p$. For very small particles β tends to infinity and $k \rightarrow k_C$, which means a chemical reaction kinetics controlled process even at very high temperatures. For finite particle diameter the larger is the diameter, the lower is the temperature at which the combustion converts from chemical to diffusion controlled.

There are three main methods of measurement of the effective combustion rate constant of a devolatilized nonporous particle:

- on the basis of mass loss of a single particle or a sample of particles,
- by means of particle diameter variation,
- on the basis of measurement of particle ignition temperature.

The simplest and for this reason most frequently used is the method based on measurement of the particle mass loss Δm_p at a known time interval of experiment Δt. If the relative mass loss $\dfrac{\Delta m_p}{m_p}$ is smaller than 0.5, then the change of particle diameter is minor. Assuming a constant density ρ_p of the particle we can calculate the effective rate constant

$$k_m = \frac{1}{m_p} \frac{\Delta m_p}{\Delta t} \frac{1}{C_{O_2,\infty}} \qquad (2.37)$$

A lot of kinetic data obtained by this method were published for various carbonaceous particles. Investigators traditionally try to eliminate the boundary layer diffusion by application of small size particles or by high velocity flows. It is important then to notice here the two principal assumptions that strongly influence the results:

- the particle temperature is equal to the gas temperature,
- no corrections for boundary layer diffusion were made.

Field et al. (1967) published data for the rate constant in equation (2.31) developed on the basis of experiments available in the 1960's:

$$k_o = 86.2 \, \frac{kg}{m^2 s(kPa)}, \qquad E = 149.5 \, \frac{kJ}{mol},$$

which can be used in a modified form of equation (2.27)

$$\dot{R}_C = k_C A_p p_{O_2, d_p} \qquad (2.38)$$

The data by Field et al. (1967) have gained wide confidence and are commonly recommended in literature for general carbon. The value of activation energy $(E = 120 - 160) \, \dfrac{kJ}{mol}$ is believed to be adequate for nonporous particles in temperature region where the combustion is controlled by chemical reaction kinetics.

The temperature difference between the particle and the surrounding gas can be calculated by means of energy balance for the particle (eq. 2.90). According to Laurendeau (1979) the particle temperature can be even 200°C higher than the gas temperature, which can significantly influence the kinetic data. Smith (1971a,b) first corrected the previously measured rate constants using the calculated particle temperature instead of the measured gas temperature and substracted the diffusion effect by means of equation (2.30).

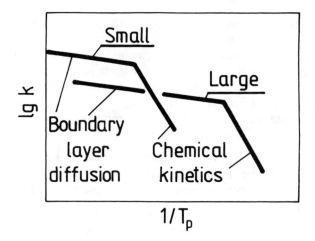

Fig. 2.15. Effective rate constant as a function of particle temperature and diameter.

Table 2.3 **Kinetic Data for Coal Combustion Rate (C + O$_2$ → CO$_2$)**

$$\dot{R}_C = k_o \cdot exp(-E/RT_p) \cdot A_p \cdot p_{O_2,\infty}^N$$

Fuel	$k_o, \dfrac{kg}{m^2 s (kPa)^N}$	$E, \dfrac{kJ}{mol}$	N	Authors
Anthracite char	1.91	79.6	1	Smith, 1970
Anthracite char	0.99	71.2	1	Smith, 1971a
Semi-Anthracite char	2.02	79.6	1	Smith, 1971b
Petroleum coke	1.98	75.4	1	Smith, 1971a
Bituminous char	0.79	67.0	1	Smith, 1971a
Subbituminous char	270	117.1	1	Tomeczek and Wójcik, 1990
Lignite char	9.24	67.0	0.5	Hamor et al.,1973
Petroleum coke	5.4	82.0	0.5	Rybak et al.,1986
Petroleum coke	7.0	82.4	0.5	Young and Smith, 1980
Petroleum coke	5.4	65.6	0.5	Mitchell and McLean, 1982
Coal char	4.0	62.7	0.5	Rybak et al.,1986
Coal char	28.0	80.2	0.5	Young, 1980
Coal char	51.2	86.5	0.5	Mitchell and McLean, 1982
Anthracite	0.081	65.3	0.5	Tomeczek and Wójcik, 1990

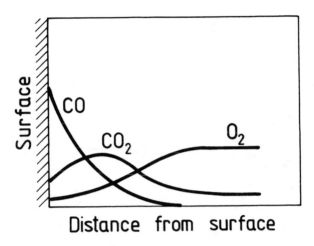

Fig. 2.16. Distribution of oxygen, carbon monoxide, and carbon dioxide.

Table 2.3. presents the effective rate constants for combustion reaction described by equation

$$\dot{R}_C = kA_p p_{O_2}^N \qquad (2.39)$$

developed by various methods. Only the data which are corrected for diffusion and particle temperature effects or are obtained by direct measurement of particle temperature (Tomeczek and Wójcik, 1990) are included in table 2.3.

The single step mechanism of combustion, during which only CO_2 is formed on the surface, is too simplified. One of the fundamental problems in combustion research was: what are the primary products of reaction CO or CO_2? It has been established that both CO and CO_2 are the primary products and the ratio CO/CO_2 increases with temperature and decreases with pressure. The ratio of CO/CO_2 produced at the solid surface as primary products increases with temperature, according to the function

$$\frac{CO}{CO_2} = A \exp(-E/RT_p) \qquad (2.40)$$

Arthur (1951) for temperatures 400–900°C presents values: $A = 2500$ and $E = 52$ kJ/mol. For the upper temperature 900°C the ratio $CO/CO_2 = 12$. If equation (2.38) was valid also for higher temperatures, then the direct formation of CO_2 on the surface is negligible. Van der Held (1961) concluded that at temperatures over 1100°C there is an equilibrium of Boudouard reaction $C + CO_2 \rightleftarrows 2CO$ at the solid surface. Consequently then, a secondary homogeneous oxidation reaction of CO takes place within the boundary layer. Figure 2.16 presents quantitatively the distribution of concentrations along the distance from the particle surface.

At high surface temperatures (>1000°C) the concentration of CO_2 at the surface can be assumed to be negligible. Carbon monoxide forming at the surface reacts with oxygen within the boundary layer. The concentration profiles, presented in figure 2.17, form two regions. In region I the concentration of oxygen $C_{O_2} = 0$, so then 2 moles of CO diffuse from the surface against 1 mole of CO_2 diffusing to the surface. This difference (2 − 1) mole generates

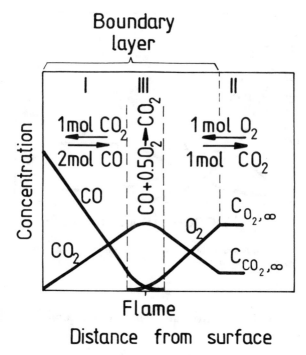

Fig. 2.17. Concentration profiles for a two step combustion model.

a net flow from the surface, called Stefan effect. In region II the concentration of carbon monoxide is equal to zero and 1 mole of O_2 diffuses to the surface against 1 mole of CO_2 diffusing to the free space. In region III a homogeneous reaction takes place with the rate

$$\dot{R}_{III} = k\,C_{CO}\,C_{O_2}^{1/2} \qquad (2.41)$$

Assuming a very narrow region III it is possible to find the radius at which the combustion of CO takes place. Jeschar and Specht (1984) calculated the radius of the flame taking into account the Stefan effect within an isothermal boundary layer.

Within region II the molar stream of oxygen is equal to the counter molar stream of carbon dioxide

$$\text{for} \quad r > \frac{d_f}{2}; \quad \dot{n}_{O_2} = -\dot{n}_{CO_2} \qquad (2.42)$$

where

$$\dot{n}_{O_2} = -4\pi\,r^2 D_{O_2}\frac{dC_{O_2}}{dr}, \quad \dot{n}_{CO_2} = -4\pi\,r^2 D_{CO_2}\frac{dC_{CO_2}}{dr} \qquad (2.43)$$

For region I molar stream of carbon monooxide is not equal the counter molar stream of carbon dioxide, but

$$\text{for} \quad r < \frac{d_f}{2}; \quad \dot{n}_{CO} = -2\dot{n}_{CO_2} \qquad (2.44)$$

This inequality of streams generates the Stefan flow, so then the streams in

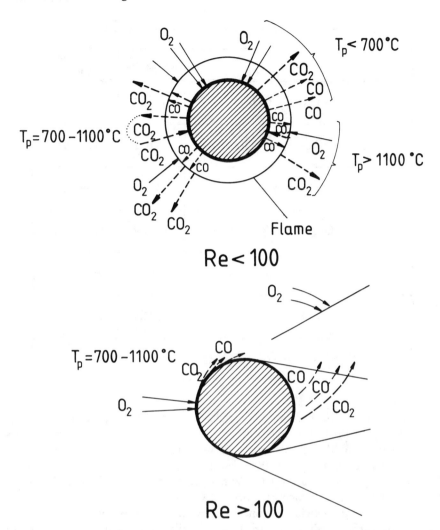

Fig. 2.18. Modes of particle combustion at various temperatures: a, stagnant particle; b, flowing particle.

equation (2.44) are equal:

$$\dot{n}_{CO} = -4\pi r^2 D_{CO} \frac{dC_{CO}}{dr} + z_{CO}(\dot{n}_{CO} + \dot{n}_{CO_2}) \tag{2.45}$$

$$\dot{n}_{CO_2} = -4\pi r^2 D_{CO_2} \frac{dC_{CO_2}}{dr} + z_{CO_2}(\dot{n}_{CO_2} + \dot{n}_{CO}) \tag{2.46}$$

Jeschar and Specht solved equations (2.42) and (2.44) with the assumption $D_{CO} = D_{CO_2} = D$ and with the boundary conditions presented in figure 2.17.

For combustion in pure oxygen they found the diameter of the sphere formed by the flame $d_f = 1.7d_p$, while for combustion in air $d_f = 2.0d_p$. For cases when there is a certain CO_2 concentration in the free space surrounding the particle, the diameter of the flame sphere is within these two values

$$1.7d_p \leqslant d_f \leqslant 2.0d_p \tag{2.47}$$

At lower temperatures <1100°C of the solid surface the mechanism of reactions is more complicated. Figure 2.18 presents three characteristic par-

Table 2.4 **Enthalpy of Reactions at 0.101 MPa, 298.15 K.**

Reaction	ΔH,kJ/mol	Type of reaction
$C(s) + O_2(g) \rightarrow CO_2(g)$	-393.7	exothermic
$C(s) + \frac{1}{2}O_2(g) \rightarrow CO(g)$	-110.1	exothermic
$C(s) + CO_2(g) \rightarrow 2CO(g)$	+172.6	endothermic
$C(s) + H_2O(g) \rightarrow CO(g) + H_2(g)$	+131.4	endothermic
$CO(g) + \frac{1}{2}O_2(g) \rightarrow CO_2(g)$	-283.6	exothermic
$H_2(g) + \frac{1}{2}O_2(g) \rightarrow H_2O(g)$	-242.4	exothermic
$CO(g) + H_2O(g) \rightarrow CO_2(g) + H_2(g)$	-41.2	exothermic

ticle temperatures for a stagnant particle and for a flowing particle. At temperatures $T_p < 700°C$ the reaction with oxygen takes place mainly at the surface leading to direct formation of both CO and CO_2. In intermediate temperatures $700°C \leq T_p \leq 1100°C$ both oxygen and carbon dioxide react at the surface producing mainly CO. At temperatures $T_p > 1100°C$ only carbon dioxide reacts at the surface producing CO.

For a flowing particle the concentration profile is asymmetrical. Oxygen reaches the surface only at the front side, while at the rear side the reaction with carbon dioxide takes place.

In practical cases there is always a certain amount of water vapour. The overall effective reaction taking place at the surface and in its neighbourhood:

$$\text{heterogeneous} \quad C + O_2 \rightarrow CO_2 \quad (2.48)$$
$$2C + O_2 \rightarrow 2CO \quad (2.49)$$
$$C + CO_2 \rightarrow 2CO \quad (2.50)$$
$$C + H_2O \rightarrow CO + H_2 \quad (2.51)$$
$$\text{homogeneous} \quad CO + \frac{1}{2}O_2 \rightleftarrows CO_2 \quad (2.52)$$
$$H_2 + \frac{1}{2}O_2 \rightleftarrows H_2O \quad (2.53)$$
$$CO + H_2O \rightleftarrows CO_2 + H_2 \quad (2.54)$$

The thermal effects of these reactions are presented in table 2.4. as enthalpies of reactions at standard conditions.

2.7. COMBUSTION OF A POROUS COAL PARTICLE

Char as well as coal particles are porous. Chemical reactions between gases and solid surface take place both on the outer and on the inner surfaces of

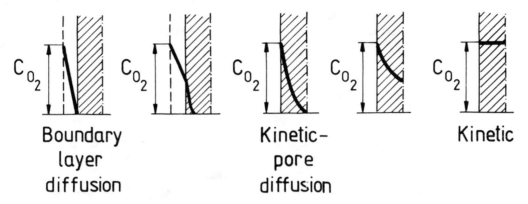

C_{O_2} C_{O_2} C_{O_2} C_{O_2} C_{O_2}

Boundary   **Kinetic–**   **Kinetic**
layer   **pore**
diffusion   **diffusion**

Fig. 2.19. Depth of oxygen penetration into porous particle for three characteristic cases.

the particle. Reacting gases must diffuse from the free space first to the particle outer surface and then through the pores structure inside the particle before they become adsorbed on the active centres. Subsequently the desorbed combustion products flow through the pores to the outer surface and then through the boundary layer to the free space.

The depth of oxygen penetration into the porous particle depends on the particle temperature. Figure 2.19 presents the three characteristic oxygen profiles:

- For high particle temperature the heterogeneous chemical reaction is so fast that oxygen is consumed at the outer surface and the process is controlled by diffusion of oxygen through the boundary layer.
- For medium temperature oxygen can penetrate deep into the particle due to slower heterogeneous reaction, but its concentration within the centre is very low. Combustion takes place both at the outer surface and nonuniformly within the particle. Process is controlled by both the chemical reaction kinetics and diffusion of oxygen within the pores.
- For low particle temperature the rate of heterogeneous chemical reaction is so slow that oxygen easily diffuses into the particle. The particle reacts within the whole volume due to a uniform oxygen concentration. Combustion is controlled only by chemical reaction kinetics.

As the combustion reaction proceeds, the size of the available pores increases, which enlarges the inner particle surface. The particle active surface reaches a maximum at burnout of about 40%, beyond which there is a decrease of the surface as a result of connection of enlarging neighboring pores.

Wheeler (1951) treats the particle as a homogeneous system of cylindrical independent pores, each with medium length L_x and diameter d_x. The number of pore mouth per unit mass of the particle, characterized by an outside surface porosity θ_s, Wheeler calculated by equation

$$N_x = \frac{A_p \theta_s}{\frac{\pi d_p^2}{4} \sqrt{2}\, m_p} \qquad (2.55)$$

where the square root $\sqrt{2}$ in the denominator is a consequence of assumption of a 45° angle between the axis of pores and the outer surface. That means that the pores' mounths have eliptical shapes.

The length and the diameter of pores characterized by internal volume V_x and internal surface area A_x are given by equations

$$L_x = \frac{V_x \sqrt{2}}{A_p \theta_s} \qquad (2.56)$$

$$d_x = \frac{4V_x}{A_x} \qquad (2.57)$$

Transport of gases within the porous particle depends on the porosity of the particle and on the diameters of pores. We can distinguish two cases:

a. For diameters of pores much bigger than the mean free path of diffusing gaseous molecules, the frequency of collisions between the gaseous molecules exceeds the frequency of their collisions with the walls of pores. Consequently the mechanism of gas transport along the pores is identical with the molecular diffusion for which the stream of i-th component diffusing through j-th component is equal

$$-D_{ij} \quad \text{grad} \quad C_i \qquad (2.58)$$

However, the mean free path depends on the molecular mass of the diffusing gas; de Boer (1953) developed for most gases a proximate equation

$$L_g = 10^{-7} \left(\frac{T}{T_n}\right) \left(\frac{p_n}{p}\right), \quad m \qquad (2.59)$$

b. For diameters of pores much smaller than the mean free path of the diffusing gaseous molecules, the frequency of collisions between the gaseous molecules and the walls of pores is much bigger than the frequency of collisions among the gaseous molecules. The gaseous molecule hitting the wall is temporarily adsorbed and then desorbed in incidental directions. This kind of mass transport is called Knudsen diffusion. The stream of i-th component diffusing through j-th is described also by diffusion equation (2.58) in which the molecular diffusion coefficient should be replaced by a Knudsen diffusion coefficient

$$D_{Ki} = \frac{2}{3} d_x \left(\frac{2RT}{\pi M_i}\right)^{1/2} \qquad (2.60)$$

Because collisions between gaseous molecules are negligible, then the streams of Knudsen diffusion for individual components are independent and each of them diffuses as if it were alone within the pores.

In the intermediate case, both mechanisms influence the diffusion rate, and the effective diffusion coefficient can be calculated by equation

$$\frac{1}{D} = \frac{1}{D_{Ki}} + \frac{1}{D_{ij}} \qquad (2.61)$$

In real coal particles pores do not form cylindrical channels of constant cross-section parallel to the diffusion direction, so the diffusing stream is smaller than that described by equation (2.58). The tortuosity of pores can be taken into account by correction of the diffusion coefficient given by equation (2.61) dividing it by the mean value of tortuosity equal 3.

The contribution of pores in reactivity of the particle can be calculated if the distribution of reacting gases is known. The balance of i-th component diffusing and reacting according to first order reaction within a pore pre-

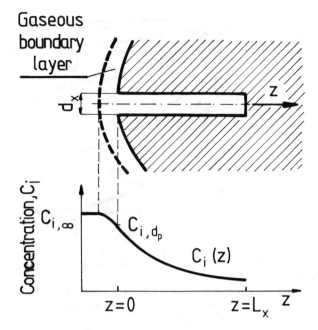

Fig. 2.20. Concentration profile of i-th gaseous component diffusing into a cylindrical pore.

sented in figure 2.20, is given by equation

$$\frac{1}{d_x^2}\frac{d}{dz}\left(d_x^2 D\frac{dC_i}{dz}\right) - \frac{4}{d_x}kC_i = 0 \qquad (2.62)$$

The boundary conditions are:

for $z = 0$

$$-D\frac{dC_i}{dz}\bigg|_{z=0} = \beta(C_{i,\infty} - C_{i,d_p}) \qquad (2.63)$$

for $z = L_x$

$$\frac{dC_i}{dz} = 0 \qquad (2.64)$$

Wheeler (1951) solved the boundary problem for the first order reaction within the pores under condition $C_{i,\infty} = C_{i,d_p}$. Assuming dimensionless values: $\Psi = \dfrac{C_i(z)}{C_{i,\infty}}$, $\xi = \dfrac{z}{L_x}$ and introducing the Thiel number

$$\mathrm{Th} = 2L_x\left(\frac{k}{d_x D}\right)^{1/2} \qquad (2.65)$$

equations (2.62)–(2.64) can be written:

$$\frac{d^2\psi}{d\xi^2} - (\mathrm{Th})^2\psi = 0 \qquad (2.66)$$

$$\text{for} \quad \xi = 0, \quad \psi = 1, \qquad (2.67)$$

$$\text{for} \quad \xi = 1, \quad \frac{d\psi}{d\xi} = 0. \qquad (2.68)$$

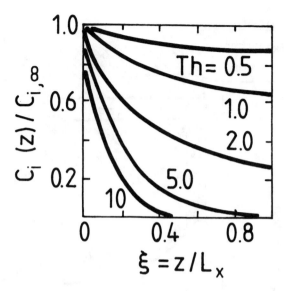

Fig. 2.21. Concentration profile of i-th gaseous component diffusing into a pore as function of Thiel number.

The solution has a form

$$\psi(\xi) = \frac{\cosh\left[(\mathrm{Th})(\xi - 1)\right]}{\cosh(\mathrm{Th})} \tag{2.69}$$

presented in figure 2.21.

The ratio of the medium concentration of i-th reacting component within the pores to the concentration on the outer surface of the particle determines the effectivity of pores

$$\eta = \int_0^1 \psi(\xi)d\xi = \frac{1}{\mathrm{Th}}\, tgh(\mathrm{Th}) \tag{2.70}$$

The rate of reaction within one pore is equal $\pi\, d_x L_x k\, C_{i,\mathrm{medium}}$, which multiplied by the number of pore mouths $N_x m_p$, gives the reaction rate of the particle

$$\dot{R} = \pi\, d_x L_x k\, \eta\, C_{i,\infty} N_x m_p \tag{2.71}$$

Application of this equation is possible if the pore structure and the kinetics constant of combustion reaction related to inner particle surface area are known. By means of figure 2.21 we can notice that for $\mathrm{Th} < 0.5$ the effectivity of pores $\eta \approx 1$. In such cases the decrease of oxygen concentration within the pores is negligible and the combustion reaction is almost uniform in the whole volume of the particle. The process is controlled by chemical reaction kinetics. For large Thiel numbers $\mathrm{Th} > 5$ oxygen does not diffuse to the ends of pores. The combustion process is then controlled by both the chemical kinetics and the oxygen diffusion within the pores. Figure 2.22. presents the oxygen concentration distribution along the pores for three characteristic cases. Because $tgh\,(5) = 0.9999$ we get $\eta = \dfrac{1}{\mathrm{Th}}$ and the rate of the

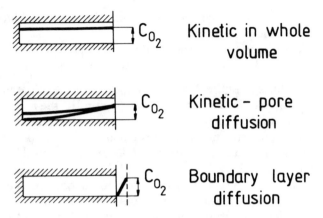

Fig. 2.22. Concentration of oxygen along the particle pore for three characteristic cases.

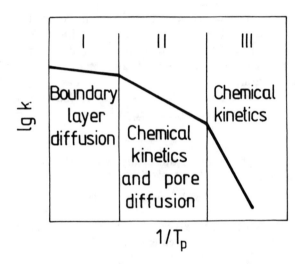

Fig. 2.23. Rate constant as a function of particle temperature.

first order reaction can be written

$$\dot{R} = \pi\, d_x^{3/2}\, N_x m_p (kD)^{1/2} C_{i,\infty} \qquad (2.72)$$

Equation (2.72) does allow us to understand the mechanism of porous particle combustion. Because the kinetic constant of chemical heterogeneous reaction is equal $k = k_o \exp(-E/RT_p)$, then at the temperature level where both the chemical kinetics and the diffusion of gases within the pores control the combustion process, the "apparent" activation energy is equal $E/2$.

For a porous particle combustion there are then three characteristic regions presented in figure 2.23:

Region I— process is controlled by oxygen diffusion from the free space to the outer surface of the particle through the boundary layer.

Region II— process is controlled by both the chemical kinetics and by the pore diffusion.

Region III— process takes place within the whole volume of the particle and is controlled by chemical kinetics.

The slope of the lines in figure 2.23 is an indicator of the apparent activation energy. For the lowest particle temperatures (region III) the activation energy is equal E, while for the intermediate temperatures (region II) the slope is twice smaller and the activation energy is equal $E/2$. In region I the slope of the line results from the temperature dependence of oxygen diffusion coefficient equation (2.36).

It is now possible to analyze the activation energies presented in table 2.3. which are, but one, within the range 60–80 kJ/mol. Comparing these values with that for nonporous particle 149.5 kJ/mol by Field et al. (1967) we come to the conclusion that the combustion of most pulverized coal particles is controlled by both the chemical kinetics and the diffusion of oxygen within the pores of the particle.

The oxidation reaction is very fast. We can expect then, that oxygen cannot diffuse deep into the particle. To calculate the depth of oxygen penetration by equation (2.69) we need information about the pores structure. There are however experimental methods allowing us to determine the mode of particle combustion.

For a porous particle burning on its outer surface we describe the rate of N-th order combustion reaction by equation

$$\dot{R} = k A_p C_{O_2,\infty}^N \tag{2.73}$$

while if the reaction takes place uniformly within the whole volume of the particle we can write

$$\dot{R} = k_m m_p C_{O_2,\infty}^N \tag{2.74}$$

The relation between the above two rate constants k and k_m for a particle with shape factor φ is given by means of equation

$$k = k_m \frac{\rho_p d_p}{6\varphi} \tag{2.75}$$

If the particle burns as a shrinking sphere the value k is independent of particle diameter, while for the particle burning with constant diameter mode (burning within the whole volume) k is inversely proportional to the diameter. The dependence or independence of the appropriate rate constant has been used first by Smith (1971a,b) as an indicator of the mode of combustion. Smith found that in pulverized coal flames particles burn almost as shrinking spheres and the rate of combustion can be determined on the basis of the particle outer surface area. Tomeczek and Remarczyk (1986) using the same procedure confirmed it also for larger particles up to 3 mm burning in fluidized beds.

2.8. COMBUSTION TIME OF A DEVOLATILIZED PARTICLE

For practical reasons combustion of coal is divided into three stages:

- heating to ignition temperature,

- combustion of volatiles,
- char combustion.

The combustion time is usually treated as a sum of the duration times of these stages. Let us consider now only the final stage, the combustion of a devolatilized particle in isothermal conditions. There are two possible extreme models of particle combustion:

- the combustion reaction takes place at the outside particle surface, so that the solid body has a constant density during the whole process (ρ_p = const),
- the particle combusts uniformly within the whole solid body, so during the process the diameter can be taken to be constant (d_p = const).

A real porous particle combusts both at decreasing diameter and density ($\rho_p \neq$ const, $d_p \neq$ const).

At constant density the rate of particle diameter variation with time of combustion can be found by means of the mass balance equation

$$\frac{d(d_p)}{dt} = -\frac{2M_C}{\rho_p} k \, C_{O_2,\infty} \tag{2.76}$$

where the rate constant for the effective $C + O_2 \rightarrow CO_2$ reaction is given by equation (2.30). Integrating equation (2.76) at constant k we get a combustion time

$$t_c = \frac{\rho_p}{2M_C \, C_{O_2,\infty}} \left(\frac{d_{po}}{k_C} + \frac{d_{po}^2}{4D} \right) \tag{2.77}$$

It should be noticed that the combustion time is a linear function of particle initial diameter when the process is kinetic controlled and a square function in diffusion controlled conditions. For very small initial diameters ($d_{po}^2 \ll d_{po}$) combustion is always controlled by chemical kinetics.

Experimental investigations at high temperatures confirm the diffusion character on the outside surface indicating a square function of the initial particle diameter. Essenhigh and Thring (1958) and Essenhigh (1961) observed a square function relation between the combustion time and the particle initial diameter for 300–4000 μm particles. Ivanova and Babij (1966) for particles 200–800 μm, within 1200–1600 K found also a square relation $t_c \sim d_{po}^2$. Tomeczek and Heród (1079) for char particles 0.2–1.4 mm and temperatures 800–1000°C measured the combustion time for two microlithotypes of subbituminous coal, presented in figures 2.24 and 2.25. A kinetic controlled region is observed up to 0.4 mm diameters. Within the whole diameter region a medium power factor $t_c \sim d_{po}^{1.3}$ has been calculated. Basu (1977) noticed during combustion of char in a fluidized bed the value of power factor $N = 1.45 - 1.78$.

2.9. COMBUSTION OF VOLATILES

The combustion of volatiles can take place close to the surface of the particle, immediately after they are ejected from the porous structure, or at some distance from the surface. In the first case the rate of combustion is controlled by the rate of pyrolysis, while in the second by the rate of mixing with the combustion air. In intermediate case the volatiles combustion rate is con-

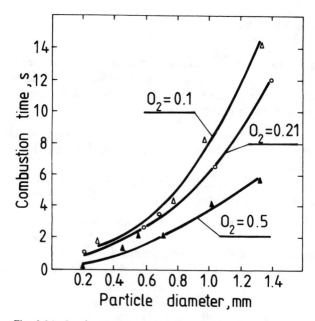

Fig. 2.24. Combustion time of vitrite particle of temperature $T_p = 1000°C$ in air with three oxygen fractions (Tomeczek and Heród, 1979).

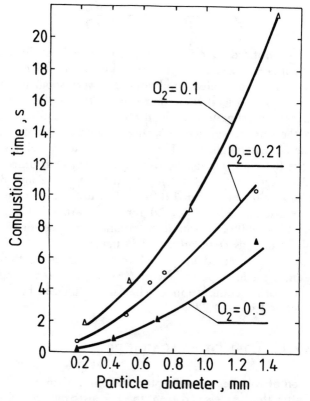

Fig. 2.25. Combustion time of inertite particle of temperature $T_p = 1000°C$ in air with three oxygen fractions (Tomeczek and Heród, 1979).

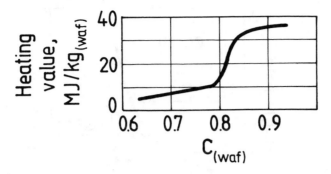

Fig. 2.26. Heating value (lower) of volatiles as a function of carbon content in coal (Wróblewska, 1978).

trolled by the kinetics of homogeneous reactions influenced by the composition of volatiles.

In most cases the volatiles mix with air during their release from the particle. If the temperature is sufficiently high, ignition will be spontaneous; however, if the temperature is not high enough, then the mixture of air and volatiles can burn as a premixed gaseous flame even at a large distance from the particle.

The properties of volatiles depend strongly on the coal type and on the conditions of their formation. The largest mass share of volatiles form tar (in gaseous state) and water. The variation of chemical composition with devolatilization conditions of such complicated substances as tar is so far not solved. There are, however, no difficulties to calculate the composition of main gaseous volatiles. The heating value of volatiles does increase with the degree of carbonization. Figure 2.26 presents the heating value (lower) as a function of $C_{(waf)}$. For lignites the heating value is small at the level of about 10 MJ/kg(waf), while for anthracite it can reach a value of 35 MJ/kg(waf).

Modelling of volatiles combustion is usually assumed to be a two step process: oxidation to carbon monooxide with subsequent oxidation to carbon dioxide. For the main products of devolatilization presented in table 2.2, the following homogeneous reactions are considered:

$$CH_4 + 1.5O_2 \xrightarrow{k_1} CO + 2H_2O \tag{2.78}$$

$$C_2H_6 + 2.5O_2 \xrightarrow{k_2} 2CO + 3H_2O \tag{2.79}$$

$$C_6H_6 + 4.5 O_2 \xrightarrow{k_3} 6CO + 3H_2O \tag{2.80}$$

$$CO + 0.5 O_2 \underset{k_{4b}}{\overset{k_{4f}}{\rightleftarrows}} CO_2 \tag{2.81}$$

The rate of the above equations can be determined by means of the Westbrook and Dryer (1981) kinetic data:

$$\dot{R}_1 = 2.8 \circ 10^9 \exp(-202.64/RT)C_{CH_4}^{-0.3} C_{O_2}^{1.3} \tag{2.82}$$

$$\dot{R}_2 = 7.31 \circ 10^9 \exp(-125.604/RT)C_{C_2H_6}^{0.1} C_{O_2}^{1.65} \tag{2.83}$$

$$\dot{R}_3 = 1.35 \circ 10^9 \exp(-125.604/RT)C_{C_6H_6}^{-0.1} C_{O_2}^{1.85} \tag{2.84}$$

$$\dot{R}_{4,f} = 2.238 \circ 10^{12} \exp(-167.472/RT)C_{CO}^1 C_{O_2}^{0.25} C_{H_2O}^{0.5} \tag{2.85}$$

$$\dot{R}_{4,b} = 5 \circ 10^8 \exp(-167.472/RT)C_{CO_2}^1 \tag{2.86}$$

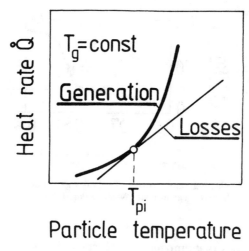

Fig. 2.27. Ignition point.

Equations (2.82)–(2.86) give the reaction rates in kmol/m³s if the activation energy is inserted in kJ/mol and the concentrations in kmol/m³.

The rate of carbon monoxide combustion reaction (2.81) in the presence of water steam is strongly influenced by H_2O concentration, even for quite small concentrations. Equations (2.85) and (2.86) have been developed by Dryer and Glassman (1972).

2.10. IGNITION OF A COAL PARTICLE

In low temperature air a coal particle reacts with oxygen, but the rate of oxidation is very slow. Consequently then the temperature of the particle does not differ much from the environment temperature. In higher temperatures the rate of oxidation increases, so that above a certain value of temperature a step-like change from slow oxidation to combustion can occur. Usually in literature the minimum value of air temperature at which the combustion initiates is called an ignition temperature. In a similar way an extinction temperature is defined.

2.10.1. Semenov analysis

A prior condition of ignition is that the rate of heat generation due to combustion must be bigger than the rate of heat losses by the particle. The particle ignition temperature T_{pi} is commonly determined by the Semenov model, the basis of which is presented in figure 2.27. According to Semenov, at the ignition point the rate of heat generation and the rate of heat losses are equal and tangent to each other at the ignition point. This analysis gives a simple solution for T_{pi} if accumulation of heat within the particle is neglected, no swelling of the particle is considered, and no volatiles release from the particle prior to ignition occurs. For this reason the simplified solution can lead to more accurate values of the ignition conditions for char and low volatile coals than for high volatile coals.

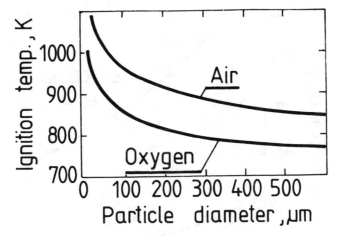

Fig. 2.28. Ignition temperature as a function of char particle diameter in air and oxygen (Rybak, 1981).

The Semenov ignition conditions can be written in the form of equations:

$$\dot{Q}_G(T_{pi}) = \dot{Q}_L(T_{pi}) \qquad (2.87)$$

$$\left.\frac{d\dot{Q}_G}{dT}\right|_{T_{pi}} = \left.\frac{d\dot{Q}_L}{dT}\right|_{T_{pi}} \qquad (2.88)$$

Heat losses due to radiation and convection for cases where the gas temperature is equal to the wall temperature, we can write in a form

$$\dot{Q}_L = A_p \left[\alpha(T_p - T_g) - \varepsilon_{p-w}(T_p^4 - T_g^4)\right] \qquad (2.89)$$

Assuming that the heterogeneous oxidation reaction, with the heat of combustion H_l, occurs at the surface of the particle we can write equation (2.87) in a form

$$A_p k_o \exp(-E/RT_{pi})C_{O_2,\infty}^N H_l = A_p \left[\frac{2\lambda_g}{d_p}(T_{pi} - T_g) + \varepsilon_{p-w}\,\sigma(T_{pi}^4 - T_g^4)\right] \qquad (2.90)$$

in which the convection heat transfer coefficient has been eliminated by utilizing the Nusselt number for small particles $Nu = \dfrac{\alpha\, d_p}{\lambda_g} = 2$. Differentiating equation (2.90) we get

$$\frac{E k_o C_{O_2,\infty}^N H_l}{RT_{pi}^2} \exp(-E/RT_{pi}) = \frac{2\lambda_g}{d_p} + 4\sigma\varepsilon_{p-w}T_{pi}^3 \qquad (2.91)$$

The last equation does allow us to calculate the ignition temperature if the kinetic constants are known. Figure 2.28 presents the value of the ignition temperature obtained numerically by Rybak (1981) as a function of particle diameter and oxygen content in the particle environment. The ignition temperature decreases with increasing particle diameter and oxygen content.

Equation (2.91) can be used also for determination of the three kinetic constants: E, k_o, and N if the ignition temperatures for particles of different

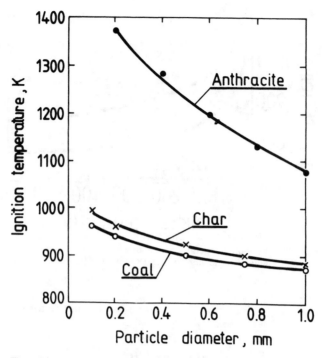

Fig. 2.29. Measured ignition temperature as a function of particle diameter (Tomeczek and Wójcik, 1990).

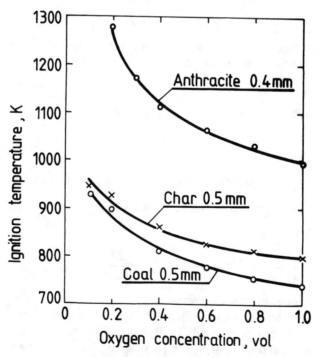

Fig. 2.30. Measured ignition temperature as a function of oxygen concentration (Tomeczek and Wójcik, 1990).

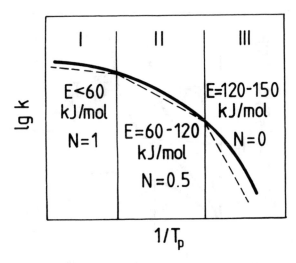

Fig. 2.31. Theoretical kinetic data for effective combustion
reaction.

diameters in different oxygen concentrations are known. Figures 2.29 and
2.30 present the values measured by Tomeczek and Wójcik (1990) of ignition
temperature for anthracite and char of subbituminous coal Siersza which were
used for kinetic constant evaluation. The activation energy can be calculated
on the basis of experiments at constant oxygen concentration, while the reaction order can be calculated on the basis of experiments at constant particle
diameter. The preexponential factor can be finally found if the former E and
N are known. The kinetic constants calculated by Tomeczek and Wójcik for
anthracite and char of subbituminous coal Siersza are presented in table 2.3.
The analysis of the activation energies together with the reaction orders on
the basis of theoretical kinetic data presented in figure 2.31 indicates that the
ignition of anthracite particle is controlled by chemical kinetics plus pore
diffusion, zone II type ($N = 1$, $E = 60$–120 kJ/mol) at the lower limit of
zone III because the reaction order is consistent very well with the theoretically expected value for zone I ($N = 0.5$). The ignition of char particle is also
controlled by chemical kinetics plus pore diffusion, zone II type but at the
upper limit.

For coal particles the evaluated value $N = 1.65$ indicates that ignition
of coal containing high amounts of volatiles cannot be described by the simple
heterogeneous Semenov mechanism.

2.10.2. Ignition time

A particle injected into an oxidizing environment of temperature higher than
the ignition temperature will ignite after a certain time, commonly called the
ignition time. This time can be determined by means of the energy balance
equation containing energy accumulation terms, because the steady-state Semenov analysis can not be applied. Assuming negligible temperature varia-
tions within the particle, chemical combustion reaction only at the outer
particle surface, and the temperature of the chamber walls equal to the gas

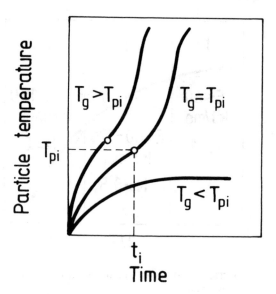

Fig. 2.32. Temperature history of coal particle at three
gas temperatures.

temperature, we can write the energy equation in a form

$$\frac{2}{3} d_p \rho_p c_p \frac{dT_p}{dt} = k_o \exp(-E/RT_p) C_{O_2,\infty} H_l$$
$$- \alpha(T_p - T_g) - \varepsilon_{p-w} \sigma(T_p^4 - T_g^4) \quad (2.92)$$

The temperature history of the particle at various gas temperatures is presented qualitatively in figure 2.32. For temperatures of the surrounding gas (usually air) smaller than the ignition temperature, the particle will heat up due to convection and radiation but also due to slow chemical combustion reaction. The particle will not ignite, but it will react with oxygen, and if left for sufficiently long time in the chamber it will burn slowly. At gas temperature equal or higher than the smallest ignition temperature, the particle temperature variation gets acceleration at ignition point. The presence of an inflexion point on the temperature curve is the condition of ignition. So then at $t = t_i$ we can write

$$\frac{d^2 T_p}{dt^2} = 0 \quad (2.93)$$

Rybak solved the energy equation (2.92) with condition (2.93), and the obtained ignition times are presented in figure 2.33 for three gas temperatures. A strong influence of the gas temperature is seen. Decreasing the gas temperature from 2023 K to 1023 K lengthens the ignition time from 6 ms to 96 ms for a coal particle of 50 μm diameter.

It must be remembered, however, that the real ignition time of a coal particle will be longer, because there will be some delay of ignition due to particle devolatilization.

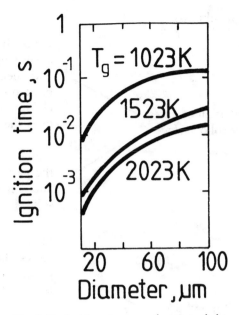

Fig. 2.33. Ignition time as a function of char particle diameter and gas temperature (Rybak, 1981).

2.10.3. General model of coal particle ignition

During heating of coal particles the devolatilization preceeds the ignition. Mixing of volatiles with the surrounding air can lead to formation of an ignitable gaseous solution. The rate of homogeneous chemical reactions within this solution depends not only on the temperature but also on the concentration of volatiles and oxygen. The increase of chemical reactions rate causes a local increase of temperature and consequently a fall of heat transferred to this local volume from surrounding gases. A moment in which no heat is transferred to the reacting volume indicates the ignition time. At this moment there is a reverse of the heat transfer from surroundings to the local volume onto from the analyzed volume to the surroundings. Assuming symmetrical devolatilization of a particle, the ignited gases form a burning sphere of diameter d_f surrounding the particle. The place at which ignition takes place can then be identified on the basis of zero value of spatial temperature gradient

$$\left. \frac{dT_g}{dr} \right|_{r=\frac{d_f}{2}} = 0 \qquad (2.94)$$

Ignition of gases can occur at the particle surface or within the bulk of gases. In literature they are called heterogeneous and homogeneous ignition. The burning gases form a flame front which shifts during combustion of the particle.

It is important to notice that when the ignition of the solid particle surface preceeds the ignition of gases we always call it heterogeneous ignition. But when the ignition of gases occurs before the ignition of the solid surface, we can have four possibilities:

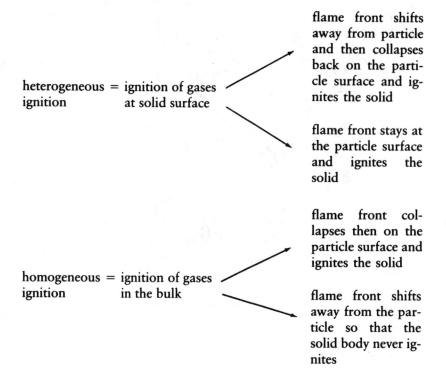

heterogeneous = ignition of gases
ignition   at solid surface

→ flame front shifts away from particle and then collapses back on the parti-cle surface and ig-nites the solid

→ flame front stays at the particle surface and ignites the solid

homogeneous = ignition of gases
ignition   in the bulk

→ flame front col-lapses then on the particle surface and ignites the solid

→ flame front shifts away from the par-ticle so that the solid body never ig-nites

There is also possible a simultaneous ignition of both the solid body and the gaseous phase.

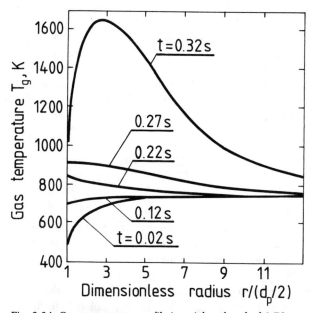

Fig. 2.34. Gas temperature profile in neigbourhood of 0.75mm particle placed in air of temperature 750K in a chamber with wall temperature 1750K (Wójcik, 1991).

Determination of the mode of ignition can be made on the basis of solution of the energy and mass balance equations within the particle and in gaeseous phase surrounding the particle:

within the particle—equations (2.13)–(2.16),
    in the neighbourhood of the particle—gaseous phase continuity equation, gaseous phase energy equation.

Wójcik (1991) solved the above set of equations. Figure 2.34 presents an example of a homogeneous ignition. Coal particle of diameter 0.75 mm is placed in air of temperature 750 K within a chamber where the walls have temperature 1750 K. The particle heats up mainly due to radiation from the walls. Ignition of volatiles occurs at the solid surface at time $t_i = 0.27$ s. The position of the flame front after ignition shifts away from the particle surface and within the analyzed time does not collapse on the solid surface.

REFERENCES TO CHAPTER 2

Andrussov L.—Z. Elektrochem., 54(1950)566–571.

Arendt P.—Dissertation TH Aachen, 1980.

Arendt T., Kaiser M., Wanzl W. and van Heek K. H.—Internationale Kohlenwissenschaftliche Tagung, Proceedings, Düsseldorf 1981.

Arthur J. R.—Trans. Faraday Soc., 47(1951)164–178.

Badzioch S. and Hawksley F. G. W.—Ind. Eng. Process Des. Dev., 9 (1970)521.

Basu P.—Fuel, 56(1977)390.

Bliek A., van Poelje W. M., van Swaaij W. P. M. and van Beckum F. P. H.—AICHE J., 31(1985)1666–1681.

Carslaw H. S. and Jaeger J. C.—Conduction of heat in solids. Clarendon Press, Oxford, 1959.

Davies W. B. and Brown D. J.—Brennstoffchemie, Bd.50(1969) 163–167.

De Boer J. H.—The dynamical character of adsorption. Oxford University Press. New York, 1953.

Dryer F. L. and Glassman I.—14th Symp. (Int.) on Combustion. The Comb. Institute, Pittsburgh, 1972.

Essenhigh R. H.—J. Inst. Fuel, 34(1961)239.

Essenhigh R. H.—J. Eng. Power, 85(1963)182.

Essenhigh R. H. and Thring M. W.—Proc. Inst. Fuel Conf. Science in Use of Coal, Sheffield D-21, 1958.

Field M. A., Gill D. W., Morgan B. B., and Hawksley P. G. W.—Combustion of Pulverized Coal. BCURA, Leatherhead 1967.

Frossling N.—Gerlands Beitr. Geophys., 52, 1/2(1938)170.

Gavalas G. R.—Coal Pyrolysis. Elsevier, N. York, 1982.

Gavalas G. R. and Wilks K. A.—AICHE J., 26 (1980)201–212.

Hamor R. J., Smith I. W. and Tyler R. J.—Combustion and Flame, 21 (1973)153.

Howard J. B. and Essenhigh R. H.—Ind. Eng. Chem. Proc. Des. Dev., 6(1967)74–84.

Ivanova I. P. and Babij V. L.—Teploenergietika, 13(1966) 54–59.

Jeschar R. and Specht E.—Über den Abbrandverlauf einzelner Kohlenstoffpartikel. Abhandlungen der Braunschweigischen Wissenschaftlichen Gesellschaft. Band XXXVI, 1984.

Jung K.—M. Eng. Sci. Thesis, Dep. of Chem. Eng., Univ. Melbourne, 1980.

Kaiser M.—Studienarbeit, Universitat Essen GHS, 1981.

Kallend A. S. and Nettleton M. A.—Erdöl und Kohle-Erdgas-Petrochem., 19, Mai (1966) 354.

Koch W., Jüntgen H. and Peters W.—Brennstoffchemie, 50(1969)369–373.

Kowol J.—PhD Thesis. Silesian Technical University, Gliwice, 1985.

Laurendeau N. H.—Progress in Energy and Combustion Science, 4(1979) 221–270.

Löwenthal G.—Doktor Arbeit, Bergbau-Forschung GmbH, Essen, 1984.

Mitchell R. E. and Mc Lean W. J.—19th Symp. (Int.) on Combustion. The Comb. Inst. Pittsburgh, 1982.

Morris J. P. and Keairns D. L.—Thesis, Dep. of Chem. Eng., Univ. Melbourne, 1980.

Niksa S., Heyd L. E., Russel W. B. and Seville D. A.—20th Symp. (Int.) on Combustion. The Comb. Inst., Pittsburgh, 1984.

Peters W.—Habilitationsschrift, TH Aachen, 1963.

Peters W. and Bertling H.—Fuel, 44(1965)317–331.

Pottgiesser C—Dissertation, TH Aachen, 1980.

Ragland K. W. and Weiss C. A.—Energy, 4(1979)341.

Rowe P. N., Claxton K. T. and Lewis J. B.—Trans. Instn. Chem. Engrs., 43(1965)31.

Rybak W.—PhD Thesis. Politechnika Wroclawska, 1981.

Rybak W., Zembrzuski M. and Smith I. W.—21st Symp. (Int.) on Combustion. The Comb. Inst. Pittsburgh, 1986.

Schwandtner D.—PhD Thesis, TH Aachen, 1971.

Simons G. A.—Combustion and Flame, 55(1984)181–194.

Simons G. A. and Lewis P. F.—Mass transport and heterogeneous reactions in a porous medium. The Comb. Inst. Spring Meeting, Central States Section, Cleveland, OH, 1977.

Smith I. W.—Combustion and Flame, 17(1971)303–314,a.

Smith I. W.—Combustion and Flame, 17(1971)421–428,b.

Smith I. W.—CSIRO Div. Mineral Chem., Invest. Report 86, December, 1970.

Spokes G. N. and Benson S. W.—Oxidation of a thin film of carbonaseous solid at pressures bellow 10^{-4} Tor, in Fundamentals of Gas-Surface Reactions. Academic Press, New York, 1967.

Stahlherm D., Jüntgen H. and Peters W.—Erdöl, Kohle-Erdgas-Petrochem., 27 February (1974)64.

Street P. J., Weight R. P. and Lightman P.—Fuel, 48(1969)343.

Suuberg E. M., Peters W. A. and Howard J. B.—17th Symp.(Int.) on Combustion. The Comb. Inst. Pittsburgh, 1978.

Tomeczek J. and Heród A.—Arch. Term. Spal., 10 (1979)4,609–616.

Tomeczek J. and Kowol J.—Canadian J. Chem. Eng., 69(1991)286–293.

Tomeczek J. and Remarczyk L.—Canadian J. Chem. Eng., 64(1986) 871–874.

Tomeczek J. and Wójcik J.—23rd Symp. (Int.) on Combustion. The Comb. Inst. Pittsburgh, 1990.

Van der Held E. F. M.—Chemical Engineering Science, (1961)300–312.

Van Heek K. H.—Druckpyrolyse von Steinkohlen. Habilitationsschrift. Univ. Münster, 1981.

Van Loon W.—Devergassing van koostof met zuurstof en stoom. Thesis, Delft 1952.

Westbrook C. K. and Dryer F. L.—Comb. Sci. Techn., 27(1981)31.

Wheeler A.—Advances in Catalysis, 3(1951)249.

Wójcik J.—PhD Thesis, Silesian Technical Univ., Gliwice, 1991.

Wróblewska V.—Arch. Term. Spal., 9(1978)623–632.

Young B.C.—Proc. Int. Conf. on Coal Sci., Düsseldorf, 1980.

Young B.C. and Smith I. W.—18th Symp. (Int.) on Combustion. The Comb. Inst. Pittsburgh, 1980.

3

PULVERIZED COAL COMBUSTION

The combustion of pulverized coal is the main way of coal utilization for heat and power generation. The severe impact of this kind of coal combustion on the environment inspired the development of coal combustion in fluidized bed. It must be however clearly stated that the recent progress in sulphur and nitrogen oxides removal in pulverized flame boilers firmly strengthened the competetive position of this technology. Figure 3.1. presents a typical geometry of a pulverized coal fired boiler.

3.1. COMBUSTION GASODYNAMICS

A fluid flowing out of a nozzle forms a jet acting dynamically on its surroundings. The cross-section of the nozzle is usually much smaller than the size of the combustion chamber, so the influence of the chamber walls on the development of the jet can be neglected.

A jet spreads along its path, losing velocity and entraining gas from the surroundings. In a free jet the static pressure is constant, while in a confined jet the pressure increases with the distance from the nozzle causing formation of a recirculation zone. Temperature differences between the jet and the surroundings create bouyancy forces, which can curve the jet.

In technical solutions pulverized coal is transported by part of the combustion air. This part, usually called primary air, amounts to about 20–30% of the combustion air. The primary jet is commonly surrounded by a jet of the secondary air, but in some solutions the secondary air nozzles are removed to some distance from the primary air nozzle. Sometimes the stream of the total air is divided into three parts, so a nozzle of tertiary air is used.

In pulverized coal flames solid particles are dispersed in gaseous phase. The velocity of the solid particles and gas are not equal. What is more, the gravity forces act also on particles, but because the fall velocity of pulverized coal ($<100 \mu$m) is very small this influence can not be significant. The inertia forces, however, due to change of direction of the gas flow can considerably influence the flame.

The intensity of recirculation of hot gases to the initial part of the jet plays an important role in stabilization of the flames. An outer recirculation zone is formed in a confined straight jet, while an inner recirculation zone can be generated by a bluff body placed in the main stream of the jet or by a strong swirling of the jet initiated by the burner nozzles.

The velocity at the outflow of nozzles in modern furnaces is high, so that the jets have turbulent character. Although most practical cases are three dimensional, a lot of useful information can be obtained from simpler cases of two dimensional jets.

3.1.1. Straight free jets

A jet in a free space is formed of four parts presented in figure 3.2.:

- core of the jet,
- mixing region,

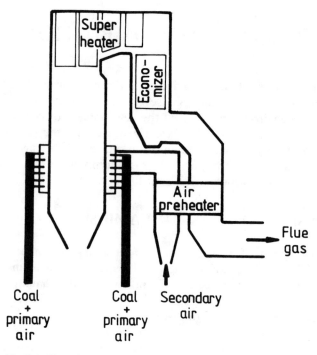

Fig. 3.1. Typical geometry of a pulverized coal boiler.

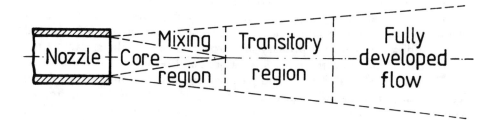

Fig. 3.2. Characteristic regions of a free straight jet.

- transitory region,
- fully developed jet.

The core region is characterized by constant velocity and concentration of fluid equal to that at the nozzle. Outside this region develops a boundary layer in which momentum and mass are transported perpendicular to the direction of flow. The length of the core region is equal to about 4–5 nozzle diameters (or nozzle widths in the case of plane jets). The transitory region, which can reach a distance of 10 nozzle diameters from the source, precedes the fully developed jet.

The fully developed regions of turbulent jets are similar; consequently then the axial and radial distribution of velocity and concentration can be described by general equations. The profiles of concentration and temperature in jets are essentially identical, which confirms the analogy of heat and mass transfer.

The shape of the jet is characterized by the jet half-angle, which is contained between the jet axis and the line formed by points in which the velocity is equal a half of the velocity in the axis. The average value of the jet half-angle is equal:

$$\text{plane jet} \quad - \quad \varphi_w = 5.5°,$$
$$\text{round jet} \quad - \quad \varphi_w = 4.85°.$$

In a similar way the concentration profile within the jet is defined for which the half angle is equal:

$$\text{plane jet} \quad - \quad \varphi_C = 7.9°,$$
$$\text{round jet} \quad - \quad \varphi_C = 6.2°.$$

A jet flowing out of a nozzle of diameter d_o with density ρ_o at the nozzle outlet, develops in the fluid of density ρ_a surrounding the jet. In the fully developed region of the jet the velocity and concentration profiles can be described by equations:

Axial profiles—

velocity,

$$\frac{w_x(x,0)}{w_o} = k_{w,x} \left(\frac{\rho_o}{\rho_a}\right)^{1/2} \left(\frac{d_o}{x}\right)^N \tag{3.1}$$

concentration,

$$\frac{C(x,0)}{C_o} = k_{C,x} \left(\frac{\rho_o}{\rho_a}\right)^{1/2} \left(\frac{d_o}{x}\right)^N \tag{3.2}$$

where the power factor $N = 1/2$ for a plane jet and $N = 1$ for a round jet. Van der Hegge Zijnen (1958) stated a better agreement with the experimental data if in the above equations the origin of x coordinate is shifted. He proposed to substitute $(x + 0.6\, d_o)$ in the velocity profile and $(x + 0.8\, d_o)$ in the concentration profiles instead of x.

Transversal profiles—

velocity,

$$\frac{w_x(x,r)}{w_x(x,0)} = \exp\left(-k_{w,r}\left(\frac{r}{x}\right)^2\right) \tag{3.3}$$

or   (3.4)

$$\frac{w_x(x,r)}{w_x(x,0)} = 0.5\left(1 + \cos\left(\frac{\pi r}{\alpha_w x}\right)\right)$$

concentration,

$$\frac{C(x,r)}{C(x,0)} = \exp\left(-k_{C,r}\left(\frac{r}{x}\right)^2\right) \tag{3.5}$$

or   (3.6)

$$\frac{C(x,r)}{C(x,0)} = 0.5\left(1 + \cos\left(\frac{\pi r}{\alpha_C x}\right)\right)$$

For a plane jet in the above equations (3.1)–(3.6) the nozzle diameter d_o should be replaced by the nozzle width h_o and radius r by coordinate y perpendicular to jet axis x. The coefficients k_w, k_C, α_w and α_C according to Van der Hegge Zijnen (1958) are presented in table 3.1.

Table 3.1 Coefficients k_w, k_C, α_w and α_C (Van der Hegge Zijnen, 1958).

Coefficient	Geometry	
	Plane	Round
$k_{w,x}$	2.48	6.3
$k_{C,x}$	2.00	5.0
$k_{w,r}$	75	96
$k_{C,r}$	36.6	57.5
α_w	0.192	0.170
α_C	0.279	0.217

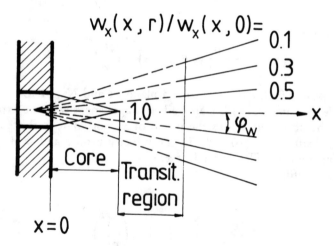

Fig. 3.3. Radial profile of relative velocity.

Figure 3.3. presents the lines of constant ratio of velocity to the axial velocity $\dfrac{w_x(x,r)}{w_x(x,0)}$ along the jet. All the lines of constant value of this ratio converge in point $x = -0.6\ d_o$. Figure 3.3. illustrates also the half-angle φ_w of the jet. The ratio of the local velocity to the nozzle velocity is presented in figure 3.4. Within the core region the velocity is constant and equal w_o, while in the transitory region the velocity decreases to about 0.8 w_o.

Entrainment of fluid by the jet along its path can be calculated according to Ricou and Spalding (1961). The stream of mass at distance x is equal

$$\dot{m}_x = \int\limits_{-\infty}^{+\infty} \rho w \, dA \qquad (3.7)$$

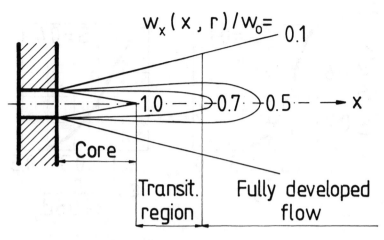

Fig. 3.4. Lines of constant velocity ratio.

where $dA = dy$ for a plane jet and $dA = 2\pi r\, dr$ for a round jet. Assuming a constant density ρ and using equation (3.3) we can write the last equation in a form

$$\dot{m}_x = \rho w(x,0) \int_{-\infty}^{+\infty} \exp\left(-k_{w,r}\left(\frac{r}{x}\right)^2\right) dA \qquad (3.8)$$

Because the mass flow at the nozzle outlet is equal $\dot{m}_o = A_o \rho_o w_o$, then after integration we get

$$\frac{\dot{m}_x}{\dot{m}_o} = 0.508 \left(\frac{\rho_a}{\rho_o}\right)^{1/2} \left(\frac{x}{h_o}\right)^{1/2} \qquad (3.9)$$

for a plane jet flowing out of a nozzle with width h_o, and

$$\frac{\dot{m}_x}{\dot{m}_o} = 0.32 \left(\frac{\rho_a}{\rho_o}\right)^{1/2} \frac{x}{d_o} \qquad (3.10)$$

for a round jet.

The above equations (3.9) and (3.10) can be successfully applied for Reynold number at the nozzle $>2.5 \circ 10^4$ and at distances $\dfrac{x}{d_o} > 6$ from the nozzle.

3.1.2. Swirling free jets

Swirling of jets by burner nozzles is applied as an effective way of flame stabilization. In case of very strong swirling a backward axial pressure gradient is formed, creating an inner recirculation flow in the centre of the jet. If the recirculation zone reaches the flame region, then the recircled high temperature combustion products ignite the fuel-air mixture causing stabilization. This way of stabilization is called an inner stabilization.

Intensity of swirling is characterized by the swirling number

$$S = \frac{2 G_\varphi}{d_o G_x} \qquad (3.11)$$

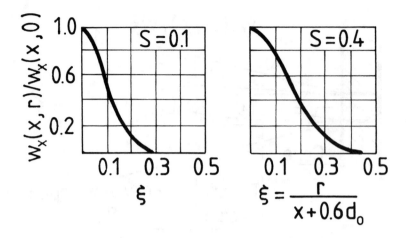

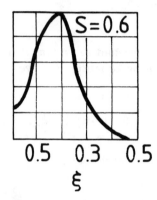

Fig. 3.5. Axial velocity profile of a swirling jet.

where the angular momentum

$$G_\varphi = \int\limits_0^{(d_0/2)} (w_\varphi r)\rho w_x 2\pi r\, dr \qquad (3.12)$$

and the axial momentum

$$G_x = \int\limits_0^{(d_0/2)} \rho w_x^2 2\pi r\, dr \qquad (3.13)$$

The swirling number does not characterize fully the jet; nevertheless it is successfully used for jets classification:

$$S \leqslant 0.2 \quad — \text{ weak swirling,}$$
$$0.2 < S < 0.6 \quad — \text{ moderate swirling,}$$
$$0.6 \leqslant S \quad — \text{ strong swirling.}$$

Weak swirling is not strong enough to generate a reverse axial gradient in the centre of the jet in order to create internal recirculation. Figure 3.5. presents the axial component of the velocity. Two kinds of profiles are clearly seen. For $S < 0.5$ the profile is described by the Gaussian distribution, while

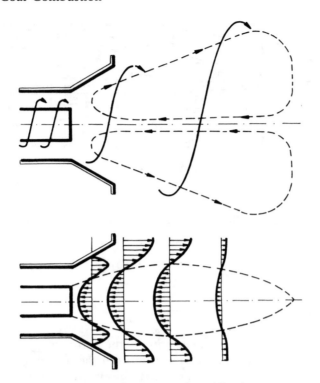

Fig. 3.6. Inner recirculation zone of a swirling jet.

for $S > 0.5$ the maximum axial velocity is not in the axis. Recirculating flow in the centre appears for swirling degree $S > 0.6$.

In case of weak and moderate swirling a similarity of velocity profiles develops already at $\frac{x}{d_o} \geq 2$, that is much nearer than in the straight jets. The axial velocity within the similarity region can be described by equation

$$\frac{w_x(x,r)}{w_x(x,0)} = \exp\left(-\frac{92}{1 + 6S}\left(\frac{r}{x + 0.6d_o}\right)^2\right) \qquad (3.14)$$

The axial velocity along the jet axis for weak swirling and $x > 4\,d_o$ decreases according to a formula

$$\frac{w_x(x,0)}{w_o} = \frac{6.8}{1 + 6.8S^2}\frac{d_o}{x + 0.6d_o} \qquad (3.15)$$

A half-angle of the jet, defined on the basis of points $\frac{w_x(x,r)}{w_x(x,0)} = 0.5$ does increase with the swirling degree. For a weak swirling Chigier and Chervinski (1967) obtained a relation

$$\varphi_w = 4.8 + 14\,S \qquad (3.16)$$

Strong swirling does increase the half-angle. Also the shape of the outflow nozzle has influence on the recirculation zone. Chedaille et al. (1966) have shown that the most favourable half-angle of the nozzle outlet is equal to $35°$ and its length $(1-2)d_o$. Figure 3.6. presents the recirculation zone for one shape of the nozzle.

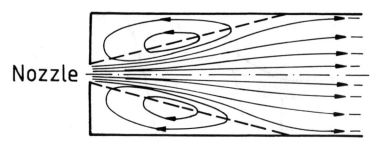

Fig. 3.7. General profile of a confined jet.

Entrainment of fluid by the free swirling isothermal jet can be calculated by equation

$$\frac{\dot{m}_x}{\dot{m}_o} = (0.32 + 0.8S)\frac{x}{d_o} \qquad (3.17)$$

3.1.3. Confined jets

A jet in an enclosed space cannot entrain the surrounding fluid. Pressure within the jet increases along the distance from the nozzle in contrast to a constant pressure in a free jet. There are, however, not sufficient data indicating that a confined jet does change its half-angle. Figure 3.7. presents characteristic velocity distributions within a confined jet. Close to the wall there is a backward flow, forming an outer recirculation region.

Thring and Newby (1953) proposed a simple theoretical treatment of recirculation based on empirical equations allowing us to calculate the entrainment of a free jet. This theory assumes that a jet preserves the properties of a free jet until a point (C), placed in the middle of a distance between the point $x = \dfrac{d_o}{0.32}\left(\dfrac{\rho_o}{\rho_a}\right)^{1/2}$ in which entrainment is equal to zero and the point (P) in which the jet contacts with the walls. This assumption is valid only in cases when the effective nozzle diameter is small in comparison with the chamber width (or diameter).

For a round jet the half-angle is $\varphi_w = 4.85°$ and the distance from the nozzle to the point where the jet contacts the wall of the chamber of width L_k (or diameter), is equal

$$x_P = 2.925L_k \qquad (3.18)$$

The stream of fluid entraining the jet between points $x = 0$ and

$$x_C = \frac{1}{2}\left(x_P + \frac{d_o}{0.32}\left(\frac{\rho_o}{\rho_a}\right)^{1/2}\right) \qquad (3.19)$$

is equal

$$\dot{m}_r = 0.32\left(\frac{\rho_a}{\rho_o}\right)^{1/2}\frac{x_C}{d_o}\dot{m}_o - \dot{m}_o = \left(0.32\frac{x_C}{d_e} - 1\right)\dot{m}_o \qquad (3.20)$$

where the effective nozzle diameter is described by equation

$$d_e = \frac{2\dot{m}_o}{(\pi\rho_a G_o)^{1/2}} = d_o\left(\frac{\rho_o}{\rho_a}\right)^{1/2} \qquad (3.21)$$

The concept of an effective nozzle diameter can be applied also to multi co-flowing jets, allowing us to replace them by a single jet with a mass stream $\dot{m}_o$ and momentum G_o equal to the sum of the values of individual jets. Introducing a dimensionless Thring-Newby parameter

$$\Theta = \frac{d_o}{L_k}\left(\frac{\rho_o}{\rho_a}\right)^{1/2} \tag{3.22}$$

we get

$$x_C = L_k\left(2.925 + \frac{\Theta}{0.32}\right) \tag{3.23}$$

and a recirculation ratio (Field et al., 1967)

$$\frac{\dot{m}_r}{\dot{m}_o} = \frac{0.32\,x_C}{L_k\Theta} - 1 = \frac{0.47}{\Theta} - 0.5 \tag{3.24}$$

The last equation is valid for round jets and for small $d_e/L_k \leqslant 0.05$. For $d_e/L_k > 0.1$ the theory of Cray and Curtet (1955) can be applied. This theory is based on more general assumptions, but creates difficulties in practical applications.

In free jets the axial velocity decreases to zero with the distance from the nozzle, while in confined jets it tends to a limited value. Beyond the point where the jet contacts the wall the radial velocity profile flattens so that at distance $2L_k$ they reach a profile characteristic for turbulent flow in channels.

3.1.4. Curved jets

There are two common cases when the axis of the jet does not form a straight line. First, the combustion chambers are not isothermal; that can cause jet curvature. Second, the jet can flow out the nozzle into a cross-flowing stream. In both cases the shape of the jet has to be determined.

Intensive heat transfer from combustion gases to the cooling surfaces causes large differences of temperature between the flame and its surroundings. Consequently then the buoyancy forces generated within the combustion chamber can lift the flame up because the density of the flame is smaller than that of the surroundings. The deformation of the flame axis depends on the ratio of the gravitation and inertia forces. The shape of the axis can be calculated by means of the following equation (Field et al., 1967)

$$\frac{y}{d_e} = 0.047\,\frac{\text{Ar}}{\text{Re}^2}\left(\frac{x}{d_e}\right)^3 \tag{3.25}$$

Beer et al. (1963) produced equation

$$\frac{y}{d_e} = 0.065\,\frac{\text{Ar}}{\text{Re}^2}\left(\frac{x}{d_e}\right) \tag{3.26}$$

The shape of a jet in a cross-flowing stream depends on the ratio of momentum of the jet and of the cross-flowing stream $\rho_o w_o^2/\rho_q w_q^2$. For a jet

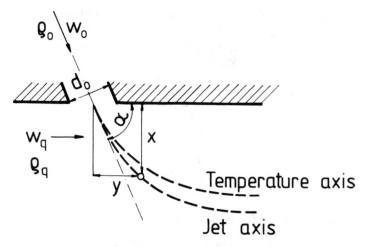

Fig. 3.8. Jet profile in a cross-flowing stream.

presented in figure 3.8 the jet axis can be described by equation

$$\frac{y}{d_o} = 0.01 \left(\frac{\rho_q w_q^2}{\rho_o w_o^2}\right)^{1.3} \left(\frac{x}{d_o}\right)^3 - 0.07 \frac{x}{d_o} tg(\alpha - 90°) \qquad (3.27)$$

Noncircular nozzles can be treated as circular with equivalent nozzle diameter, as can rectangular nozzles if the ratio of their sides' length is smaller then 30.

A distance between the nozzle and a cross-section of the jet in which the difference between the jet axis velocity and the velocity of the stream is equal to $0.05(w_o - w_q)$ is called the depth of jet penetration. For a perpendicular geometry $\alpha = 90°$ the dimensionless jet penetration can be calculated from a proximate equation

$$\frac{L_s}{d_o} = 2 \frac{w_o}{w_q} \sqrt{\frac{T_q}{T_o}} \qquad (3.28)$$

In the case of a multi-jet system, presented in figure 3.9., the shape of the jet can be described by equation (3.27) if the ratio of the chamber depth to the nozzle diameter is $B/d_o \geq 44$. For ratio $s/d_o \geq 22$ the following equation can be used

$$\frac{y}{d_o} = 0.104 \frac{\rho_q w_q^2}{\rho_o w_o^2} \left(\frac{x}{d_o}\right)^{3.25} \qquad (3.29)$$

The depth of jet penetration can be calculated by relation

$$\frac{L_s}{d_o} = K \frac{w_o}{w_q} \sqrt{\frac{\rho_o}{\rho_q}} \qquad (3.30)$$

where the value K depends on the ratio s/d_o as presented in figure 3.10.

3.1.5. Physical modelling of flames

Despite enormous progress in mathematical modelling of turbulent flames, in many practical cases the designers still have to rely on physical modelling.

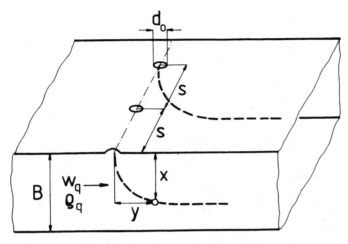

Fig. 3.9. Geometry of a multi-jet system.

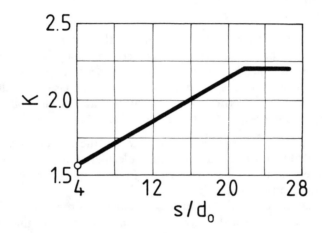

Fig. 3.10. Value K in equation (3.30).

The success of this kind of modelling depends mainly on the ability to insure a similarity of phenomena taking place in the model and in the practical system.

The simplest way of modelling is the geometrical similarity, which demands an equal ratio of appropriate linear dimensions in the model and in the practical system. The geometrical similarity must be fulfiled in the case of incompressible fluids or in isothermal conditions. In flames, however, this similarity criteria must be rejected because the density of gases varies considerably.

The general criteria of similarity can be identified either by means of dimensional analysis or by transformation of equations describing the process into dimensionless form. The first way has this disadvantage, that it can not distinguish between physical values of the same dimension (heat, enthalpy, work) and needs a lot of experience. The only reliable way is the second way of identification of dimensionless groups in balance equations of the process.

In physical modelling of combustion the following dimensionless criteria are most often used:

— Reynolds number

$$\text{Re} = \frac{\text{velocity} \circ \text{linear dimension}}{\text{kinematic viscosity}} = \frac{w\,L}{\nu} \tag{3.31}$$

— Froude number

$$\text{Fr} = \frac{(\text{velocity})^2}{\text{linear dimension} \circ \text{gravity}} = \frac{w^2}{Lg} \tag{3.32}$$

— Prandtl number

$$\text{Pr} = \frac{\text{kinematic viscosity} \circ \text{density} \circ \text{specific heat}}{\text{thermal conductivity}} = \frac{\nu\,\rho\,c_p}{\lambda} \tag{3.33}$$

— Schmidt number

$$\text{Sc} = \frac{\text{kinematic viscosity}}{\text{diffusion coefficient}} = \frac{\nu}{D} \tag{3.34}$$

— Archimedes number

$$\text{Ar} = \frac{(\text{linear dimension})^3 \circ \text{gravity} \circ \text{temp. difference}}{(\text{kinematic viscosity})^2 \; \text{temperature}} = \frac{L^3 g\,\Delta T}{\nu^2 T} \tag{3.35}$$

— Lewis number

$$\text{Le} = (\text{Pr})/(\text{Sc}) \tag{3.36}$$

In most cases it is not possible to fulfil all the criteria of similarity, and what is more it is not necessary. Through critical analysis it is possible to select those criteria which have to be fulfilled in order to model certain aspects of the process. However, it is not possible to define precise rules allowing us to detect which similarity criteria must be fulfilled at a given problem; nevertheless there are some general outlines (Beér and Chigier, 1974):

- In case of large Reynolds numbers, when the combustion process is controlled by turbulent mixing, the molecular transport can be ignored and there is no need to have equal Prandtl and Schmidt numbers in the model and in real process.
- For small ratio of the Archimedes number and the square of the Reynolds number, when the inertia forces are much larger than the buoyancy forces, there is no need to keep this ratio equal in the model and in the real process.
- The two phase flame can be modeled by a gaseous flame when the free fall velocity of the solid phase is much smaller than the velocity of the jet.
- In modelling of flames by isothermal models the geometrical similarity of burners should be rejected, and the nozzle diameter should be taken according to Thring-Newby theory (3.21).
- During modelling of heterogeneous flames the solid particles should be modelled according to the ratio of the linear scale to the velocity scale.

In most frequent cases of modelling there is a need to examine a combustion chamber by an isothermal model. The modelling procedure is based on an assumption, that the formation of a burning jet depends on the total momentum, which for a single free jet can be written in a form

$$G = 2\pi \int_0^\infty \rho w^2 r\,dr = G_o = \frac{\pi}{4} d_o^2 \rho_o w_o^2 = \frac{\pi}{4} d_e^2 \rho_a w_o^2 \tag{3.37}$$

The concept of an effective nozzle diameter has such consequences that in an isothermal model of temperature T_a with nozzle diameter d_e, with the jet ejected with the same velocity as in the real object w_o, we get both the mass flow and the momentum of the jet equal to those in the real object. From the last equation we can write

$$d_e = d_o \left(\frac{\rho_o}{\rho_a}\right)^{1/2} = d_o \left(\frac{T_a}{T_o}\right)^{1/2} \tag{3.38}$$

For a known mass flow of a free jet $\dot{m}_o$ and its momentum G_o at the nozzle outflow in the real object we can write, that an effective nozzle diameter in isothermal model of fluid density ρ_a (temperature T_a) is equal

$$d_e = \frac{2\dot{m}_o}{(G_o \, \pi \, \rho_a)^{1/2}} \tag{3.39}$$

In the case of a modelling of a jet confined in a chamber of width L_k two practical cases can be distinguished:

a. A stream $\dot{m}_o$ ejected from a nozzle d_o is surrounded by a secondary stream of air $\dot{m}_{o,1}$ with small velocity.
b. A stream $\dot{m}_o$ is surrounded by a secondary stream of air $\dot{m}_{o,1}$ with significant velocity, so that both jets have large momentum.

In case "a" the similarity criteria can be written as (Beér and Chigier, 1974)

$$\frac{\dot{m}_o + \dot{m}_{o,1}}{\dot{m}_o} \frac{d_o}{L_k} \tag{3.40}$$

For a burning jet in a chamber of temperature T_a the effective nozzle diameter should be used in the above equation

$$\frac{\dot{m}_o + \dot{m}_{o,1}}{\dot{m}_o} \frac{d_e}{L_k} \tag{3.41}$$

The last similarity criteria is called in literature a Thring-Newby recycling criteria for a confined jet which does allow us to predict the velocity profiles and the rate of mass recirculation

$$\Theta = \frac{\dot{m}_o + \dot{m}_{o,1}}{\dot{m}_o} \frac{d_e}{L_k} \tag{3.42}$$

In case "b" the jet at some distance from the burner nozzle has properties of a single homogeneous jet. The Thring-Newby criteria reduces then to a form

$$\Theta = \frac{d_e}{L_k} \tag{3.43}$$

where the effective nozzle diameter is equal

$$d_e = 2 \frac{\dot{m}_o + \dot{m}_{o,1}}{[(G_o + G_{o,1})\pi \rho_a]^{1/2}} \tag{3.44}$$

In order to model the near burner part of the jet in case "b" where the jet has heterogeneous properties, a similarity of the model and object can be obtained if equal ratio of mass streams are fulfilled

$$\left.\frac{\dot{m}_o}{\dot{m}_{o,1}}\right]_{model} = \left.\frac{\dot{m}_o}{\dot{m}_{o,1}}\right]_{object} \tag{3.45}$$

Modelling of jets with buoyancy forces is based on the Archimedes number as a similarity criteria. For a jet ejected with velocity w_o from a nozzle of diameter d_o into an environment of temperature T_a, the ratio of the Archimedes number and the square of the Reynolds number is equal

$$\frac{\mathrm{Ar}}{\mathrm{Re}^2} = \frac{d_o g (T_f - T_a)}{w_o^2 T_f} \qquad (3.46)$$

where T_f is the temperature of a flame jet.

It has been shown that buoyancy effect is significant for $\dfrac{\mathrm{Ar}}{\mathrm{Re}^2} > 0.01$ (Beér and Chigier, 1974). For average conditions of confined burning jets $\dfrac{T_f - T_a}{T_f} < 0.85$, so then we can write that the buoyancy can cause curvature of the flame axis if

$$\mathrm{Fr} = \frac{w_o^2}{g\, d_o} < 85 \qquad (3.47)$$

3.1.6. Gasodynamics of pulverized coal combustion chambers

Pulverized coal is supplied to the combustion chamber by means of a stream of air commonly called primary air. The size of the coal particles (less than 100 μm (150μm)) is limited first of all by the possibility to burn out during the particle residence time within the flame and also by the necessary conveying transport velocity of primary air within the pipes, which for 100μm particles is equal about 25 m/s. The rest of the air is supplied as a secondary air to the burner in the neighbourhood of the coal-primary air mixture. A common solution is that the primary air with fuel is ejected through the centre of the burner. In some solutions a tertiary air is also supplied in a somewhat remoted position.

The shape of the combustion chamber and the position of burners are the main factors influencing the velocity field and consequently the combustion performance. High efficiency of combustion can be achieved by breaking up the primary air-coal mixture into few streams and separating them by streams of secondary air. During development of pulverized coal combustion systems a variety of burners positions were tested. Theoretically any position is imaginable: bottom, silling, walls or corners of the chamber. The modern solutions prefer multi-burners firing due to:

- improved utilization of the combustion chamber volume,
- higher stability of flames interacting each other,
- widened regulation regions.

Although it is not possible to formulate absolute rules for burner design, nevertheless some general quides can be postulated:

- the choice of burner depends on the type of fuel,
- the flame generated by the burner must be stable within the whole regulation region,
- the flame should not contact the walls of the chamber.

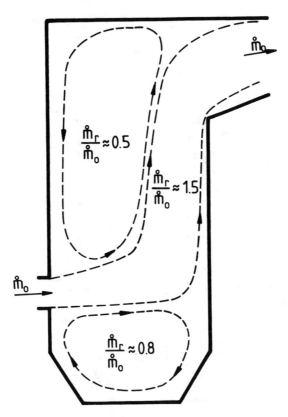

Fig. 3.11. General flow field for a one side-wall fired boiler.

One side-wall burners. Burners situated on one side-wall allow for comfortable configuration of air and coal-primary air pipes. However in such solutions there are three main disadvantages:

- low utilization of combustion chamber volume,
- poor flame stabilization,
- small burn-out,
- slagging of combustion chamber walls,
- uneven heating of walls.

Figure 3.11. presents the gasodynamics of a one side-wall firing. We can see two recirculation zones. At the bottom, below the burners, intensive recirculation results in a stream of recirculating mass equal to about 80% of the mass stream flowing in through the burner nozzles. A second upper recirculation zone (Fig. 3.11) occupies some 2/3 of the chamber cross-section, but the stream of recirculating mass is much lower $0.5 \, \dot{m}_o$ than in the bottom zone (Hzmalian and Kagan, 1976). Because the entrainment of gases is much higher from the bottom region, its existance plays a major role in flame stabilization. Tilting the burner axis at certain angle upwards (some 10 grades) reduces the bottom recirculation region. One side-wall firings have poor ability to influence the gasodynamics of the chamber. The only possibilities are through:

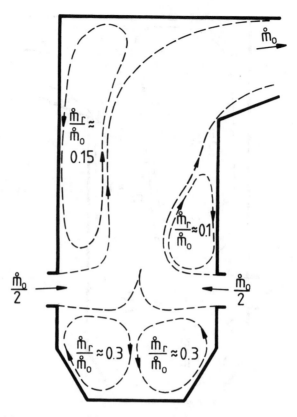

Fig. 3.12. General flow field for an opposite side-wall fired boiler.

- geometry of the burners,
- tilting of burners axis,
- velocity of air stream.

Opposite side-walls burners. Figure 3.12. presents qualitatively the velocity profiles within a chamber fired by symmetrical burners from the front and back walls. At the bottom part, below the burners two intensive recirculations zones are formed. The contact time of gases with the walls during recirculation is much shorter than in the one side-wall firing. Consequently, the temperature of gases entraining the burner jet is higher; that improves the stability of the flames. The stream of gases flowing upwards in the combustion chamber occupies about 40% of the chamber cross-section directly above the burners and expands later to some 70% (Hzmalian and Kagan, 1976). That causes higher fuel burn-out than in one side-wall firing. Heating of front and back walls is much more even. There are two possible configurations of burners:

- burners opposing each other,
- displaced burners.

It is commonly accepted that the displacement of burners situated on the front and back walls causes more even heating of the walls and better burn-

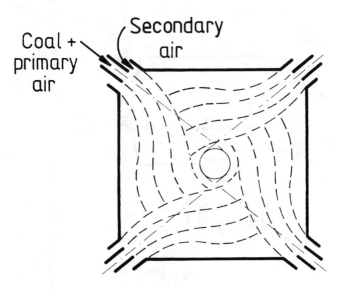

Fig. 3.13. General flow field for a tangential fired boiler.

out of fuel due to increased recirculation of high temperature gases between the displaced jets.

Corner burners. Various possibilities of corner firing can be realized, among which tangential firing is the most suitable. Compound burners placed at each corner of the combustion chamber at three to five levels at the lower third of the chamber, form in the centre of the chamber a helical flame presented in figure 3.13. In chambers of almost square cross-section the diameter of the centre circle is equal to some 10–15% of the side width. The whirling flame moves upwards and the slope of the flame axis increases. The slope of the flame and the combustion efficiency can be controlled by the tilt of the burner tip axis, which can be turned within 10–30 degrees. Below the burners a bottom whirling recirculation zone is formed.

Tangential firing leads to very stable flame. All walls of the chamber are evenly heated. The main disadvantage is, however, due to the increased slagging, particularly at the burner levels.

Cyclone burners. Injecting coal with primary air tangentially into a combustion chamber of a small size causes the centrifugal force to throw the particles on the walls of the chamber. A secondary air stream introduced also tangentially intensifies further the rotation of particles. Consequently, the combustion of coal is so intensive and the flame temperature so high that the ash particles melt. The swirling numbers for cyclone combustors is higher than for burners, $8 < S < 20$.

There are two general types of cyclone burners presented in figure 3.14.: horizontal and vertical. The horizontal burners are always slightly tilted downwards (10–15 grades) to allow the slag to flow out through the tap. In case of a cyclone cooled by water, the melted slag particles freeze during the contact with water cooled tubes on walls. The thickness of the slag layer increases during operation and its temperature rises. After this temperature reaches the liquid slag temperature, the slag starts to flow down the walls.

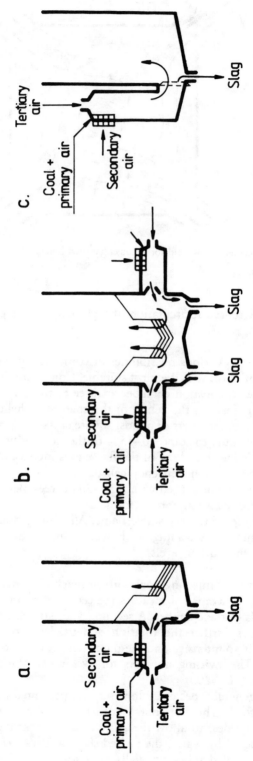

Fig. 3.14. Types of cyclone burners: a. single horizontal burner, b. multi-horizontal burners, c. vertical front burner.

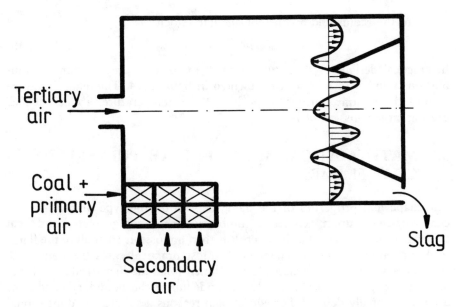

Fig. 3.15. Axial velocity profiles at cyclone burner outflow.

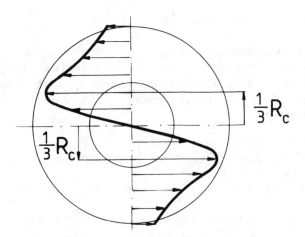

Fig. 3.16. Tangential velocity profile in a cyclone burner.

In case of a cyclone burner whose walls are not cooled, the natural refractory layer formed of slag does not develop.

Intensive rotation of flame within the cyclone, mainly due to high velocity of the secondary air 100–150 m/s, allows it to catch about 90% of ash from the flue gases within the cyclone. The length of the cyclone is equal to (1.25–1.5) of its diameter, and some 2/3 of the length occupies the secondary air nozzles. Figure 3.15 presents radial distribution of axial velocity. In the centre of the outflow duct there is a reduced pressure causing recirculation of gases from the chamber behind the cyclone. The depth of penetration of the recycling stream depends on the intensity of whirling. The tangential velocity has a maximum at radius about one third of the cyclone radius as presented in figure 3.16. Below this radius the tangential velocity is proportional to the radius and a constant angular velocity ω

$$w_\varphi = \omega r \qquad (3.48)$$

In practical designs cyclone burners most often form only a primary combustion chamber of a boiler as presented in figure 3.14. The main disadvantage of the cyclone burners is the high flame temperature, about 1800°C, causing large emission of NO_x.

3.2. MATHEMATICAL MODELLING OF PULVERIZED COAL FLAMES

The shape and properties of the flame depend on the type of coal, size of coal particles, burner geometry, combustion chamber geometry and heat transfer to the walls of the chamber. It is then impossible to analyze the flame neglecting the combustion chamber. In most practical cases the flames are three dimensional, nevertheless the two dimensional mathematical models are also able to produce useful results of modelling if the boundary conditions are professionally defined. For educational reasons a simple one dimensional model will be presented below.

3.2.1. One-dimensional model

In one-dimensional models the particles and gas flow along one geometrical coordinate with the same velocity. Such simple models can supply information about the position of ignition zone, temperature along the flame, and burnout of the particles. The balance equations will be presented under the following assumptions:

- particles do not contain volatiles,
- initial particles are divided into N size groups with $g_{o,i}$ mass fraction of i-th group,
- particles of all fractions have identical physical and chemical properties,
- particles are spherical with initial diameter $d_{po,i}$ of i-th fraction,
- particles exchange heat by convection with gas and by radiation with walls of constant temperature,
- particles burn with constant density at particle surface, according to reaction $C + \frac{1}{2}O_2 \rightarrow CO(H_1 = 9175 \text{ kJ/kg})$ while within the gaseous phase the reaction $CO + \frac{1}{2}O_2 \rightarrow CO_2(H_2 = 10129 \text{ kJ/kg})$ takes place,
- the gaseous components are O_2, CO_2 and N_2.

The problem has $2(N + 1)$ unknown parameters:

N values of burn-out of particles of each size group—b_i,
N values of particle temperature of each size fraction—$T_{p,i}$,
gas temperature—T_g,
oxygen partial pressure—p_{O_2}.

The necessary $2(N + 1)$ equations are: effective rate of combustion reaction of each particle size group, energy balance for particles, energy balance for gas, and oxygen balance.

The rate of burn-out of i-th size group of particles is equal to

$$\frac{db_i}{dt} = \frac{6 \circ 12}{\rho_p \, d_{po,i}} k(1 - b_i)^{2/3} \, C_{O_2} \tag{3.49}$$

where the rate constant k depends on particle temperature.

Energy balance for particles of i-th group has a form

$$c_p \frac{dT_{p,i}}{dt} = \frac{H_1}{1 - b_i} \frac{db_i}{dt} - \frac{12}{d_{po,i}^2 \, \rho_p} \frac{1}{(1 - b_i)^{2/3}} \lambda_g (T_{p,i} - T_g)$$
$$- \frac{6}{d_{po,i} \, \rho_p} \frac{\varepsilon_{p-w} \, \sigma}{(1 - b_i)^{1/3}} (T_{p,i}^4 - T_w^4) \tag{3.50}$$

where the first term on the right hand side means the rate of heat generation due to heterogeneous chemical reaction on the particle surface, the second means heat loss due to convection between the particle and gas, and the third term means the heat loss by the particle to the walls.

Energy balance for gas has a form

$$\frac{dT_g}{dt} = \frac{\dot{m}_{po}}{\dot{m}_g c_{pg}} \left[\sum_{i=1}^{N} g_{o,i} H_2 \frac{M_{CO}}{M_C} \frac{db_i}{dt} + 12 \frac{\lambda_g}{\rho_p} \sum_{i=1}^{N} \frac{g_{o,i}(1 - b_i)^{1/3}}{d_{po,i}^2} (T_{p,i} - T_g) \right] \tag{3.51}$$

where the first term on the right hand side means the rate of heat generation due to homogeneous chemical reaction and the second a convection heat flux from particles to gas.

The oxygen balance can be written in a form

$$\frac{dp_{O_2}}{dt} = -\frac{dp_{CO_2}}{dt} = -\frac{p}{12} \frac{\dot{m}_{po}}{\dot{m}_g} M_g \sum_{i=1}^{N} \frac{g_{o,i}}{dt} \frac{db_i}{dt} \tag{3.52}$$

where the mass stream of gas is equal

$$\dot{m}_g = \dot{m}_{go} + \sum_{i=1}^{N} \dot{m}_{po,i} \, b_i \tag{3.53}$$

It is worth noting that in one-dimensional flow it is possible to replace the time in equations (3.49) through (3.52) by a geometrical coordinate along the direction of flow

$$dx = w \, dt = \frac{\rho_{go} w_o}{\rho_g} dt \tag{3.54}$$

Mc Kenzie et al. (1974) solved the above equations for particles within the region $5-130$ μm and the results are presented in figure 3.17. The ignition zone is at a distance 2.8 m from the burner. The temperature of the small coal fractions is higher than the gas temperature by not more than 40K. The largest particles (130μm) along the whole flame have temperatures lower then the gas temperature. For wall temperature $T_w = 1400$ K a 75% burn out is achieved after 0.32s, while for a more realistic case of lower wall temperature much longer time 0.66s is needed.

One-dimensional models do not allow us to analyze the influence of recirculation of combustion products on the flame structure. Entrainment of

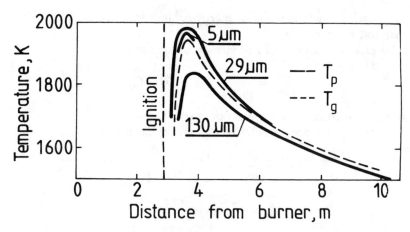

Fig. 3.17. Calculated temperature of gas and particles in a one-dimensional flame: coal particles 5–130μm, 83% below medium size 34μm (Mc Kenzie et al., 1974).

high temperature gases causes the increase of flame temperature but also reduces concentration of oxygen within the flame. Hedley and Jackson (1965) noticed an optimal recirculation degree due to these two counter effects. They included the entraining of gas into the one-dimensional model by correcting the gas stream $\dot{m}_{go}$ and its composition (Hedley and Jackson, 1966). Physically it means an assumption of infinitely fast mixing of the main jet with entraining gases. They have shown in this way, for an adiabatic combustion chamber and stechiometric air, that the optimum value of recirculation degree at which the particle combustion time is the shortest is equal to 40%.

3.2.2. Two-dimensional model

The one-dimensional pulverized flame models are seldom justified for a real furnace chamber. The computational expense of multi-dimensional models increases considerably with the number of dimensions, so then the main efforts are directed towards two-dimensional axi-symmetric models allowing their application to practical flames. The theoretical basis and solution procedures of two-dimensional homogeneous flows with recirculation were developed by Gosman et al. (1969). Gibson and Morgan (1970) extended this model to heterogeneous flows of burning particles. They obtained considerable discrepancies between the measured and computed burn-out and temperature field, which probably attributed to the over simplified models of turbulence and of coal devolatilization. Richter and Sacha (1979) using a more sophisticated model of turbulence and a temperature dependent model of coal devolatilization obtained a resonably good agreement of two-dimensional modelling with measurements.

In pulverized fuel flames there is a two-phase flow field. A detailed mathematical model should include momentum balances of both phases linked by the drag forces at the particles. Much simplification can be obtained, however, by assumption of no relative velocity between solid particles and gases at every point, which is reasonable for particles smaller than 100μm.

For a homogeneous quasi-steady flow of gas the balance equations have a form:

continuity

$$\vec{\nabla} \circ (\rho \vec{w}) = 0 \tag{3.55}$$

momentum

$$\vec{\nabla}[(\rho \vec{w}) \circ \vec{w} -] - \eta_{ef} \vec{\nabla} \left(\frac{\vec{w}^2}{2}\right) + \vec{\nabla} p = 0 \tag{3.56}$$

energy

$$\vec{\nabla} \circ \left[(\rho \vec{w})i - \lambda \vec{\nabla} T - \sum_l \rho i_l D_l \vec{\nabla} g_l - \eta_{ef} \vec{\nabla} \left(\frac{\vec{w}^2}{2}\right) \right] - \dot{q}_r = 0 \tag{3.57}$$

l-th gaseous component

$$\vec{\nabla} \circ [(\rho \vec{w})g_l - D_l \rho \vec{\nabla} g_l] - \dot{R}_l = 0 \tag{3.58}$$

Richter and Saha solved the above equations (3.55)–(3.58) assuming:

- in all equations the density ρ means the density of a gas-solid mixture, calculated as a sum of gas density and particle density,
- the effective dynamic viscosity is a sum of the laminar and turbulent viscosities

$$\eta_{ef} = \eta + \eta_t \tag{3.59}$$

where the turbulent viscosity by means of the $k - W$ model of turbulence is related to the mean value of the specific kinetic energy of turbulence k and the mean square of fluctuations of the vorticity vector W (Spalding, 1972)

$$\eta_t = \rho k W^{1/2} \tag{3.60}$$

- coal volatiles contain CH_4, CO, CO_2 and H_2O. Release of CH_4 is described by a single reaction kinetic equation (2.5), while CO, CO_2 and H_2O are released directly at the burner mouth,
- combustion of methane is according to a two step mechanism (2.78) and (2.81),
- char burning follows two overall reactions $C + \frac{1}{2}O_2 \rightarrow CO_2$ and $CO + \frac{1}{2}O_2 \rightarrow CO_2$. The heterogeneous reaction is described by a first order Arrhenius-type kinetics and the homogeneous CO combustion formed at particle surface is infinitely fast;
- no swelling of coal particles takes place,
- solid particles burning at constant density are divided into five size groups,
- the source term in an energy equation describes the net radiative heat locally absorbed per unit volume and unit time. This term is calculated with the help of a so called 4 flux radiation model, where the real exchange of radiative heat in an enclosure is approximated by 4 first-order differential equations of fluxes in the positive and negative directions of the axial and radial coordinates.

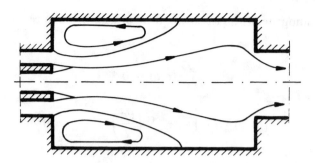

Fig. 3.18. Furnace and burner configuration.

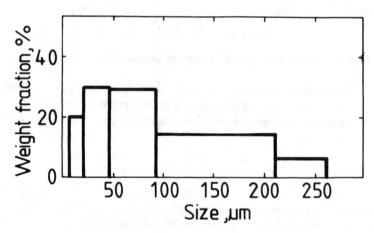

Fig. 3.19. Pulverized coal particle size distribution.

Figure 3.18 presents the geometrical configuration of a cylindrical furnace with the burner on the front wall. The boundary conditions of the outlet
of a nonswirling burner describe the velocity, temperature and concentrations
of air and of coal in the primary-air nozzle. The pressure of both fluids at
the nozzle is known. At the walls of the chamber the velocity of gases is equal
to zero. Heat is transfered to the walls by radiation and convection.

Richter and Saha (1979) solved the problem for coal particle size distribution shown in figure 3.19. Two kinds of coal were simulated: anthracite
and high volatile coal $V_{(waf)} = 37.4\%$. The burning of volatiles strongly influences the ignition behaviour of pulverized coal flames, which is revealed
by the axial temperature distribution presented in figure 3.20. The temperature rise on the axis starts earlier and the temperature gradients are steeper
in the high-volatile flame compared with the anthracite flame.

The formation of an annular zone of burning volatiles at the primary jet
edge predicted by the model is presented in figure 3.21. The volatile matter
of particles in an annular zone is released earlier and is burnt earlier than on
the axis, so the hot gases of this zone diffuse to the axis. The predicted
position of the annular zone of burning volatiles is placed further down-
stream than actually measured. This may be explained by the fact that the
mixing of the primary jet with hot recirculating products is under-predicted,

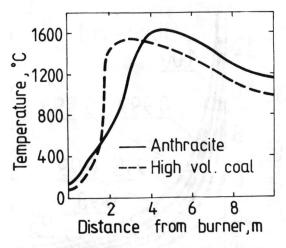

Fig. 3.20. Axial flame temperature distribution (Richter and Saha, 1979).

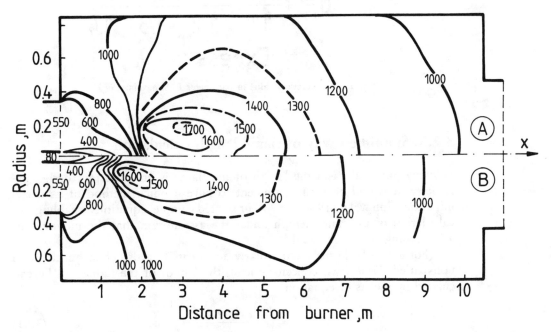

Fig. 3.21. Comparison of calculated "A" and measured "B" isotherms for high volatile coal flame (Richter and Saha, 1979).

and consequently also the temperature at the jet edge. In the second part of the furnace, the temperature profiles are mainly influenced by the burn-out of carbon or char particles and by the radiative transport of the liberated heat to the furnace walls. In this section the agreement between the predicted and measured isotherms of flames is quite satisfactory.

Further development of two-dimensional models has been made by Gosman et al. (1978) and Lockwood et al. (1980). They considered a relative velocity between the gas and the particles.

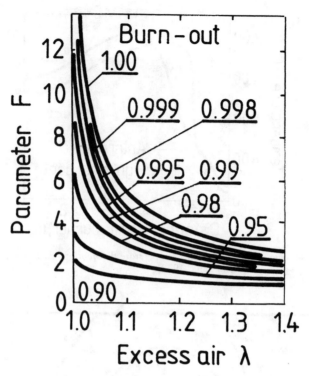

Fig. 3.22. Theoretical value of parameter F (Sunavala, 1973).

3.2.3. Simplified way of flame length calculation

In many practical cases the length of the flame is an important design parameter. It would be then convenient to have a quick method of its determination. Sunavala (1968, 1973) presented simple equations enabling the calculation of the flame length under the assumption that the mixing time and burning times are equal.

For low volatile fuels, particularly anthracite but also char, burning in a nonswirled flame (two concentric jets), the length of the flame can be calculated by means of equation

$$L_f = \left(\frac{2F_C K_C w_o d_{po} d_e}{K_w} \right)^{1/2} \tag{3.61}$$

The constant K_C is defined on the basis of combustion time $t_c = F_C K_C d_{po}$, which means a kinetic region of particle burning. Sunavala recommends a mean value $K_C = 2.1 \circ 10^4$ s/m for particles below 300 μm. The value K_w meaning the slope of axial velocity curve, which for nonswirling jet is equal $K_w = 0.147$ (Sunavala, 1973).

The function F_C depends on air excess ratio λ and volatile matter V and is defined by equation

$$F_C = \frac{\lambda}{\lambda - V} F \tag{3.62}$$

where the parameter F related to excess air ratio and particle burn-out is presented in figure 3.22. Assuming the expected burn-out degree of coal particles it is possible to calculate the length of the flame at which this burn-out value can be achieved.

Analysis of equations (3.61) and (3.62) points out that the increase of volatiles content in coal causes the increase of the flame length, in contrary to the experiments. This can be overcome for high volatile swelling coals by correction of parameter F_C in equation (3.61), which should be replaced by

$$F_C' = \frac{d_{po}}{d_{p,\exp}} F_C \qquad (3.63)$$

where $d_{p,\exp}$ is the diameter of coal particles d_{po} after swelling in standard conditions.

Swirling shortens the flame. For a swirling number $S = 1.0$ the length of the flame can be about 1/3 of the nonswirling ($S = 0$) flame length. In proximate calculations the above equations can be used if the value K_w is calculated by equation

$$K_w = 0.147 + S^2 \qquad (3.64)$$

A confuzor type burner nozzle shortens the flame by about 30%, while the impact of the secondary air at an angle 20 grades onto the central primary air-coal jet increases the value of K_w from 0.147 to 0.27 and shortens the flame by about 20%.

The above equations (3.61–3.64) can be applied for mono or narrow-sized particles. Equation (3.61) gives good agreement with experiments for coal particles burn-out not higher than 98%. In case of a deeper burn-out the calculated flame length is too short. The largest particles burn in atmosphere of very low oxygen concentration, so then the calculated length of the end part of the flame where the end part of a coal mass burns is too short. Sunavala proposes to calculate this end part of the flame by equation

$$L_{f,\,\text{end}} = \left(\frac{2 F_C' K_C w_o d_e}{K_w} \right)^{1/2} (d_{p,\max} - d_{po}) \qquad (3.65)$$

in which F_C' is determined for complete (100%) burn-out of particles having a mean initial diameter d_{po}.

3.3. MINERAL MATTER BEHAVIOUR IN FLAMES

Mineral matter is formed of two parts: the origin minerals and the external minerals. The origin minerals contain chemical elements retained from the origin plants, most frequently iron, lime, magnesium, basic metals, and sulphur. Origin minerals are equally distributed within the organic matter and usually do not exceed 4%. External minerals contain unorganic compounds from neighbouring mineral rocks. Large differences of these two mineral parts cause different behaviour during combustion.

In pulverized coal flames strong changes of mineral matter take place due to high temperature which exceeds 1600°C. The mass loss of the mineral matter in flames is higher than in standard conditions. The mineral matter introduced with coal to the combustor is received in form of:

- ash from the bottom of combustion chamber or slag in case of wet-bottom,
- slag deposits on chamber walls,

Table 3.2 **Principal Reactions by Mineral Matter During Combustion**
(Tsai, 1982; Taylor, 1976; Padia, 1977).

Species	Reaction	Temp. Range °C
Kaolinite $Al_2Si_2O_5(OH)_4$	$Al_2Si_2O_5(OH)_4 \rightarrow Al_2O_3 + 2SiO_2 + 2H_2O$	480 °C
Pyrite FeS_2	$2FeS_2 + \frac{11}{2}O_2 \rightarrow Fe_2O_3 + 4SO_2$	400÷500 °C
	$FeS_2(s) \rightarrow FeS(s) + \frac{1}{2}S_2(g)$	200÷700 °C inert. atmosph.
Sulphates $CaSO_4$	$CaSO_4 \rightarrow CaO + SO_3$	1180 °C
$MgSO_4$	$MgSO_4 \rightarrow MgO + SO_3$	1124 °C
$Fe(SO_4)_3$	$Fe(SO_4)_3 \rightarrow Fe_2O_3 + 3SO_3$	480 °C
Na_2SO_4	$Na_2SO_4(l) \rightarrow Na_2SO_4(g)$	884 °C
Carbonates $CaCO_3$	$CaCO_3 \rightarrow CaO + CO_2$	750÷850 °C
$CaMg(CO_3)_2$	$CaMg(CO_3)_2 \rightarrow CaO + MgO + 2CO_2$	730÷760 °C
Chlorides $NaCl$	$NaCl(s) \rightarrow NaCl(l)$	800 °C
	$NaCl(l) \rightarrow NaCl(g)$	1465 °C
	$NaCl(g) + H_2O(g) \rightarrow NaOH(g) + HCl(g)$	1030÷1230 °C

- fouling deposits on heating surfaces,
- fly ash.

Table 3.2. presents the principal reactions taking place in mineral matter during combustion, consequences of which are illustrated in figure 3.23. The changes of minerals are initiated by the loss of water steam, which are followed by release of CO_2, H_2O, SO_2, and SO_3 and oxidation of pyrites. Sintering starts at temperature 650°C. Above 1000°C liquid phases appear, leading to slag formation. Volatization of alkalis (Na_2O and K_2O) becomes significant at temperature above 1100°C. Gasification of SiO_2 is important above 1650°C leading to fly ash formation in oxidizing conditions. Above 2200°C only CaO and Al_2O_3 remain as a solid phase (Reid, 1981).

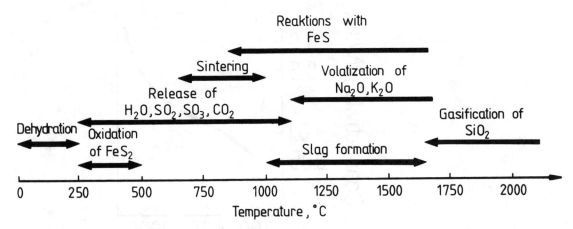

Fig. 3.23. Mineral matter transformation with temperature.

Silica reactions are responsible for fly ash formation. Losses of silica during combustion are due to formation of volatile SiO by reaction

$$SiO_2(s) + C(s) \rightarrow SiO(g) + CO(g) \qquad (3.66)$$

Padia et al. (1977) observed a 4% loss of silica at 1557°C and 20–30% loss above 1694°C due to the above reaction.

Reoxidation of SiO into SiO_2 produces fused fly ash submicron particles which can grow by adhesion and coalescence. Ulrich et al. (1977) pointed out that the submicron fly ash formed by reaction (3.66) and subsequent oxidation of SiO equals about 1% of the fly ash mass, but represents 99.5% of the number of fly ash particles.

Sodium and potassium play a main role in fouling of heating surfaces due to easy volatilization of alkalis from mineral compounds in coal. Although volatilization of alkalis is still poorly understood, it is commonly accepted that condensation of alkalis on heating surfaces is responsible for sticking of particles conveying by gases to the surfaces. The fouling index of coal ash is then determined on the basis of Na_2O content.

It is believed that NaCl is vaporized from burning coal particle and exists in flame as sodium chloride or dessociates. In addition NaCl may interact with silica or silicate particles to form sodium rich silicate particles or glazes. In high oxygen and sulphur concentration gases Na_2SO_4 is formed.

Iron silicates of low melting point in deposits show that they are released into the flame gases as products of reaction between iron compounds (Fe_2O_3 or Fe_3O_4) and silicate minerals.

Sulphur causes the main corrosion, fouling, and pollution problems. During combustion sulphur is released as SO_2 in pyrite oxidation reaction, molecular sulphur (S_2) through pyrites decomposition, or as SO_3 due to decomposition of sulphates. Organic sulphur is released in the course of complex reactions mainly in form of SO_2. Oxidation of molecular sulphur forms SO_2, part of which oxidizes later to SO_3 leading to sulphuric acid.

The deposits on the heating surfaces are formed by material conveyed to

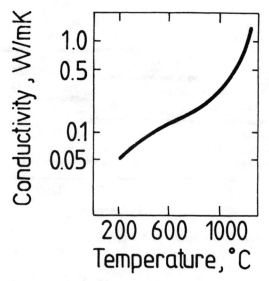

Fig. 3.24. Thermal conductivity of a medium deposit.

the surfaces by either diffusion ($<3\mu m$) or direct impact ($>10\mu m$) of particles. The mechanism of deposits formation causes its constant increase, once it is commenced. Fouling deposits occur when sulphur oxides react with alkali ash components forming alkali sulphates, which act as glue for the fly ash. The combustion chamber is too hot for solid sulphate formation. On the initial layer of deposit stick fine fly ash particles ($<3\mu m$) rich in alkalis (70%) forming the primary deposit. The growth of primary deposit is controlled by diffusion of fly ash and condensing volatiles to the heating surfaces. As the primary layer grows, the temperature of its outer surface increases. The increase of deposit surface temperature slows down the mechanism of growth and alters its nature, thus a second layer is formed. The second layer is much thicker and contains less alkalis (20%). It is believed to be formed by collection of fly ash on the sticky surface and consolidation of material by cementing and sintering. The formation of deposits proceeds until its outer surface reaches the critical viscosity temperature of ash, when the excess collected material starts to flow down the tubes. The second layer of fouling deposits can be removed from the tube surface by blowing, but that cannot be done with slag. Slagging take place when melted ash particles impact on the heating surfaces. The chemical composition of slag differs considerably from that of fly ash.

A thin layer of deposits on the heating surfaces can cause a drastic altering of heat transfer conditions, because of the small thermal conductivity of deposits. Various sources quote experimental data of thermal conductivity increasing with temperature. Figure 3.24 presents an example of thermal conductivity based on data by Boow and Goard (1969) for a medium sample of deposits. At sintering temperature (1100–1200°C) there is a sharp increase of conductivity, some (4–5) times higher than of the particulate deposit.

The thermal emissivity of deposits reported by many investigators does appear to be strongly dependent on the nature of deposits. For practical purposes a value of emissivity equal 0.6–0.7 can be assumed.

It is important to emphasize that the reason for intensive fouling or slagging is not the high content of ash in fuel but the physical and chemical properties of ash. It has to be clearly stated that so far no universal method exists allowing us to determine the slagging of a combustion chamber on the basis of chemical analysis of ash only. The burner and combustion chamber geometry have decisive influence on temperature profile within the flame and in the boundary layer of heating surfaces. If it was possible to guarantee a temperature of 800°C within the boundary layer, then no slagging would occur.

3.4. FORMATION OF NITROGEN OXIDES IN FLAMES

Emission of nitrogen oxides is responsible for some negative environmental phenomena: corrosion of metals, weakning of textiles, and slowing of vegetation. There are several nitrogen oxides, but the major pollution impacts include NO and NO_2. A notation NO_x is commonly used, which means a sum of various nitrogen oxides. Global emission of nitrogen oxides due to natural processes is much higher than that which is the product of industrial activity, however the latter concentrates highly in certain urban areas, causing danger.

It has been believed that NO_2, the directly most harmful agent, does not represent more than 5% of the emitted NO_x. Although many reports suggest much higher levels of NO_2, in some cases even 80% of NO_x, there is strong evidence that the proportion of NO_2 actually present in flames is quite small. Johnson et al. (1979) proved that the measured NO_2 concentration in the sampled gas bears little relation to the NO_2 concentration in the flame. It is notoriously difficult to measure NO_2 concentration in combustion gases, as NO and NO_2 are strongly interconverted in the sampling system, where the NO/NO_2 ratio is perturbed by reactions with flame radicals during the quenching of the sample in the sampling device.

Fenimore (1975) concluded that above 1200 K, NO_2 is in equilibrium with NO and oxygen atom

$$NO + O \rightleftarrows NO_2 \qquad (3.67)$$

Detailed kinetic calculations suggest that NO_2 formation and destruction in flames can occur by the following reactions sequence:

$$NO + HO_2 \rightleftarrows NO_2 + OH \qquad (3.68)$$

$$NO_2 + H \rightleftarrows NO + OH \qquad (3.69)$$

$$NO_2 + O \rightleftarrows NO + O_2 \qquad (3.70)$$

In low temperature regions of flames significant HO_2 concentrations are found that can react with NO formed in the high-temperature regions and transported to the low-temperature region. The NO_2 removal reactions are rapid and in the presence of high radical concentrations NO_2 will be converted rapidly back to NO.

At flame temperature NO_2 can exist only as a transient species. If NO_2 is to persist in the combustion products, then there must be quenching of NO_2 formed in the flame.

Combustion of coal in air generates nitrogen oxides partially through oxidation of air-nitrogen and partially by oxidation of nitrogen bounded chemically in coal. Formation and survival of NO_x depends mainly on thermal parameters: temperature field and oxygen concentration within the flame.

Table 3.3 **Kinetic Constants of NO Formation Mechanism (Bowman, 1975).**

Reaction	Rate Constant, $m^3/(mol \cdot s)$	Temperature Range, K
$O + N_2 \rightarrow NO + N$	$7.6 \cdot 10^7 \cdot exp(-38000/T)$	$2000 \div 5000$
$N + NO \rightarrow N_2 + O$	$1.6 \cdot 10^6$	$300 \div 5000$
$N + O_2 \rightarrow NO + O$	$6.4 \cdot 10^3 \cdot T \cdot exp(-3150/T)$	$300 \div 3000$
$O + NO \rightarrow O_2 + N$	$1.5 \cdot 10^3 \cdot T \cdot exp(-19500/T)$	$1000 \div 3000$
$N + OH \rightarrow NO + H$	$1 \cdot 10^8$	$300 \div 2500$
$H + NO \rightarrow OH + N$	$2 \cdot 10^8 \cdot exp(-23650/T)$	$2200 \div 4500$

The recorded emission of NO_x from large boilers is within 200–1500 ppm.

In pulverized coal flames the conversion of nitrogen bonds in coal into nitrogen oxides reaches 40%. It is then impossible to explain the NO_x emission on the level of several hundred ppm on the basis of coal nitrogen only. Zeldovich et al. (1947) proposed for combustion in near or over stechiometric air that formation of nitrogen oxides from molecular air-nitrogen can be explained by reactions:

$$O + N_2 \rightleftarrows NO + N \qquad\qquad (3.71)$$
$$N + O_2 \rightleftarrows NO + O \qquad\qquad (3.72)$$

Lavoie et al. (1970) noticed that reaction

$$N + OH \rightleftarrows NO + H \qquad\qquad (3.73)$$

also plays an important role. Table 3.3. presents kinetic constants of extended Zeldovich mechanism described by equations (3.71), (3.72) and (3.73).

The Zeldovich mechanism is strongly temperature dependent, so that the NO concentration depends almost entirely on the maximum flame temperature. In contrast, the conversion of coal bound nitrogen is weak temperature dependent but is a strong function of local stechiometry. Coal bound nitrogen very easily forms NO in oxidizing atmosphere, while in reducing atmosphere it can be converted into molecular N_2. In this way we come to a concept of staged combustion or recirculation of combustion products. Figure 3.25. presents a simplified mechanism allowing us to explain the reduction of NO by means of staged air supply. Coal is first devolatilized in reducing atmosphere when most of coal bound nitrogen is converted into N_2. In the next stage char is burned out in the secondary air without a risk of additional NO formation, because almost all coal bound nitrogen was previously converted into molecular N_2.

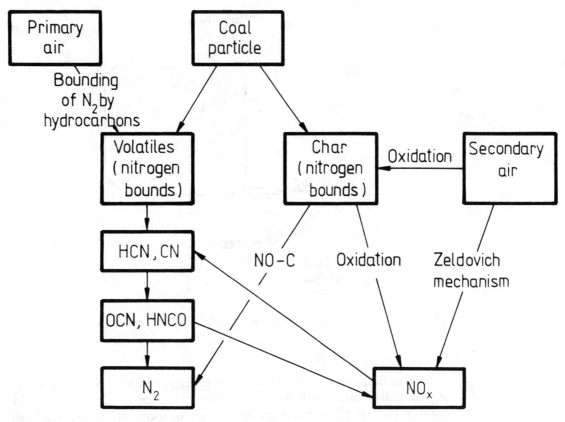

Fig. 3.25. Simplified mechanism of nitrogen oxides formation during coal combustion.

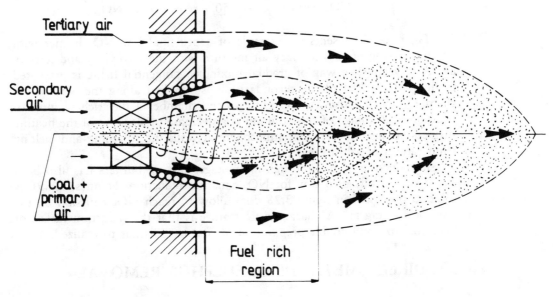

Fig. 3.26. Low NO$_x$ emission burner.

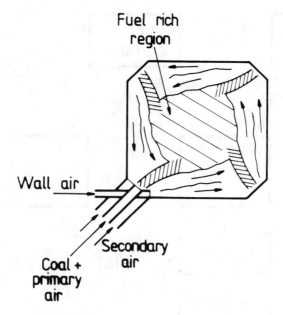

Fig. 3.27. Gradual air mixing in tangential firing.

Practical application of the staged combustion of coal for reduction of NO_x formation can be achieved by clear spatial stages of combustion or by means of burners allowing for gradual mixing of secondary air with the primary air-coal mixture. A considerable reduction of NO_x can be achieved by parallel application of various solutions:

low excess air	— 15% reduction of NO_x,
tertiary air over the burners	— 30% reduction of NO_x,
low NO_x burners	— 50% reduction of NO_x.

Figure 3.26 presents an example of a principal low NO_x burner with gradual mixing of coal-primary air mixture with the secondary and tertiary air. A very effective way of gradual mixing in tangential firing is presented in figure 3.27. Part of the secondary air is supplied along the wall of the chamber, which causes in the centre of the chamber a fuel rich region. An additional advantage of this solution is a decrease of slagging of the heating surfaces because in oxidizing atmosphere at walls the softening and melting temperatures of ash decrease.

Introduction of fuel through multi-stages mixing burners has also been tested successfully as a way for NO_x emission reduction (Hannes, 1989). A solution presented in figure 3.28 does allow to reach NO_x emission on the level of 200 mg/m³. A characteristic feature is that the reduction fuel is of finer coal size and double preparation of coal in mills has to realized.

3.5. DIRECT METHODS OF SULPHUR REMOVAL

Bounding of sulphur oxides in combustion chamber can be achieved by means of additives: CaO, $Ca(OH)_2$ and $CaCO_3$. In the case of the last two additives

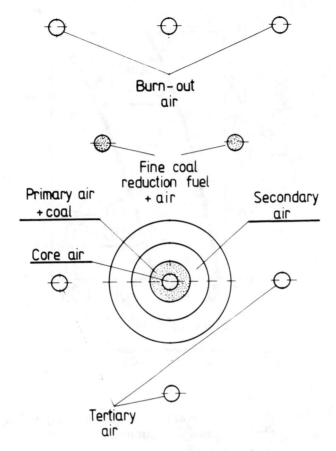

Fig. 3.28. Staged fuel and air mixing for NO_x reduction.

Table 3.4 **Enthalpy of Reactions at Standard Conditions 298.15 K, 0.1 MPa**
(Chugtai and Michelfelder, 1983)

Reaction	Enthalpy of Reaction	
	kJ/mol	kJ/kg CaO
(3.74)	+109.3	−1951.8
(3.75)	+177.4	+3167.9
(3.76)	-492	-8785.7
(3.77)	-351	-6267.9
(3.78)	-207.9	-3712.5
(3.79)	-283.7	-5066.1

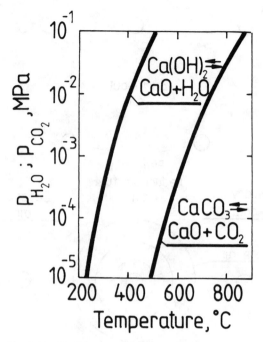

Fig. 3.29. Equilibrium curves for reactions (3.74) and (3.75) (Chugtai and Michelfelder, 1983).

the reactions of dehydration and decarbonation (calcination) within the combustion chambers precede the desulphurization:

$$Ca(OH)_2 \rightarrow CaO + H_2O \qquad (3.74)$$

$$CaCO_3 \rightarrow CaO + CO_2 \qquad (3.75)$$

Removal of sulphur oxides SO_2 and SO_3 can be obtained by means of heterogeneous reactions:

$$CaO + SO_2 + \frac{1}{2}O_2 \rightarrow CaSO_4 \qquad (3.76)$$

$$CaO + SO_3 \rightarrow CaSO_4 \qquad (3.77)$$

Parallel to reactions (3.76) and (3.77) bounding of halogens takes place:

$$CaO + 2HCl \rightarrow CaCl_2 + H_2O \qquad (3.78)$$

$$CaO + 2HF \rightarrow CaF_2 + H_2O \qquad (3.79)$$

The first two reactions are endothermic and the remaining four are exothermic, which is shown in table 3.4 on the basis of enthalpy of reactions at standard conditions.

The efficiency of direct sulphur removal in pulverized coal combustion chambers depends on the temperature level at the reaction region. High temperature favours dehydration and decarbonation, but on the other hand it leads to decomposition of the formed gypsum. Figure 3.29 presents equilibrium curves for reactions (3.74) and (3.75). The equilibrium partial pressure of H_2O and CO_2 increases with temperature. For atmospheric pressure combustion gases containing 5% H_2O and 15% CO_2 efficient dehydration can be obtained already at 300°C and decarbonation at 750°C.

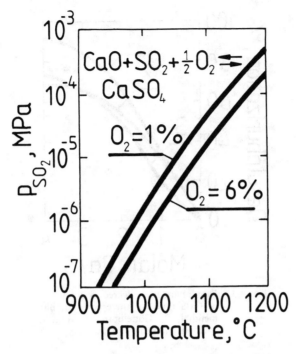

Fig. 3.30. Equilibrium curves for reaction (3.76) for two oxygen volume fractions in flue gases (Chugtai and Michelfelder, 1983).

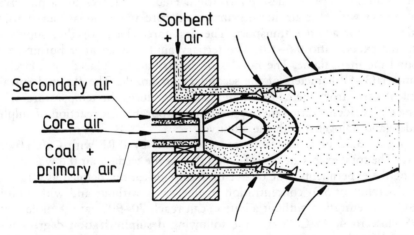

Fig. 3.31. Direct reduction of SO₂ by burner additive injection (Chugtai and Michelfelder, 1983).

Successful bounding of SO_2 according to reaction (3.76) is possible only at adequate temperature. Figure 3.30 presents the equilibrium pressure of SO_2 as a temperature function for 1% and 6% of O_2 in gases. In practical conditions then (0.1 MPa, $O_2 = 6\%$ in gases, SO_2 concentration 2340 mg/m³(STP)) the equilibrium temperature would be 1160°C. Above this temperature $CaSO_4$ decomposes.

Direct bounding of SO_2 in pulverized coal flames is possible only in a

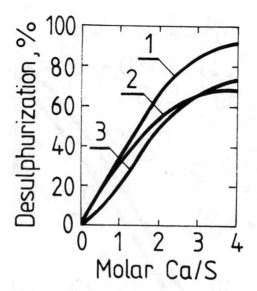

Fig. 3.32. Desulphurization degree as a function of Ca/S ratio for three coals: 1, Blumenthal, 2, Göttelborn, 3, Auguste Victoria (Flament, 1981).

narrow temperature region 750°C–1100°C, which explains why desulphurization is so difficult because the flame temperature is equal to some 1600°C. By careful organization of the mixing of additives within the flame it is possible to attain successful desulphurization. Figure 3.31 presents a pulverized coal burner with four air ducts: primary air, core air, secondary air-swirling, tertiary air for additive transport. The first three ducts together supply air with the excess ratio 0.6–1.0. The tertiary air forms an after burning zone around the main flame. The nozzles of tertiary air are placed on a circle of diameter (3–5) diameters of the secondary air nozzle. The external entrainment of gases from the combustion chamber reduces the temperature of the flame on its sides creating there conditions favourable for reaction of sulphur oxides (and halogens) with the additives.

Figure 3.32 presents results of experiments by IFRF with $Ca(OH)_2$ additive. The total excess air ratio 1.2 and that of the tertiary air 0.6 was applied during the experiments. The degree of desulphurization, defined as a ratio of the difference of SO_2 concentration in flue gases without and with additive to SO_2 concentration without additive, can reach 70–90% at the molar ratio Ca/S close to 4. At Ca/S = 2 the following desulphurization degrees were obtained for the three coals used (Chugtai and Michelfelder, 1983):

- 50%, coal "Augusta Victoria," SO_2 without additive—2385 mg/m³ (STP),
- 57%, coal "Göttelborn," SO_2 without additive—1810 mg/m³ (STP),
- 65%, coal "Blumenthal," SO_2 without additive—7712 mg/m³ (STP).

During the tests it has been noticed that introduction of additive together with the primary-coal mixture, reduces the desulphurization degree to a half of the values presented in figure 3.32.

Results of experiments with a single burner do not have to be confirmed in full industrial scale. The complex nature of the multi-burner firing influence on the temperature field within the chamber means that the position of additive supply must be found on full scale experiments. Economical calculations have shown a potential of 50% reduction of desulphurization cost by direct methods.

3.6. SECONDARY METHODS OF SULPHUR AND NITROGEN OXIDES REMOVAL

There are two principal methods of SO_2 reduction in flue gases leaving the combustion chamber:

dry methods	— adsorption of SO_2 by solids,
wet methods	— absorption of SO_2 in a water solvent of agent.

The main advantage of dry methods of desulphurization is caused by the fact that there is no need to cool the flue gases. The highest efficiency is obtained at temperatures 800–1000°C. The intensity of reaction between gas flowing through a bed of adsorbing particles depends on temperature diameter and relative velocity particle-gas. The smaller is the adsorbing particle and the bigger is the relative velocity, the more intensive is the desulphurization of flue gases. In a packed bed the limestone particles (CaO) remain a constant diameter during the process. The product of reactions (3.76) and (3.77) forms an increasing layer into the particle. Because the density of $CaSO_4$ is larger than the density of fresh sorbent, then the formation of this layer makes the diffusion of SO_2 or SO_3 into the pores more difficult. To overcome this it is necessary to apply a high ratio of Ca/S equal 2–3. Dry methods can not be utilized in processes where very high sulphur removal must be attained.

Active coke can also be used as sorbent of SO_2 but at temperatures 80–130°C. On the surface of coke the SO_2, O_2, and H_2O components of flue gases are adsorbed, reacting then with each other according to reactions (Jüntgen and Richter, 1985):

$$\left.\begin{array}{l} (SO_2)_g \rightarrow (SO_2)_{ads} \\ \frac{1}{2}(O_2)_g \rightarrow (O)_{ads} \\ (H_2O)_g \rightarrow (H_2O)_{ads} \end{array}\right\} \rightarrow (H_2SO_4)_{ads} \qquad (3.80)$$

During thermal regeneration a SO_2 rich gas is formed

$$2(H_2SO_4)_{ads} + C \rightarrow 2SO_2 + CO_2 + 2H_2O \qquad (3.81)$$

and a regenerated active coke, which can be recycled to the process.

Much better desulphurization can be reached by absorption of SO_2 in a water slurry or solution with the agent. The flue gases in the wet method must be cooled to a temperature 90–110°C. The main chemical reactions in case of lime compounds water slurry are:

$$Ca(OH)_2 + SO_2 \rightarrow CaSO_3 + H_2O \qquad (3.82)$$

$$CaCO_3 + SO_2 \rightarrow CaSO_3 + CO_2 \qquad (3.83)$$

$$CaSO_3 + \frac{1}{2}O_2 \rightarrow CaSO_4 \qquad (3.84)$$

Because the third reaction proceeds only partially, then in products both the $CaSO_3$ and $CaSO_4$ are found, together with non reacted $CaCO_3$ or $Ca(OH)_2$. As side reactions take place:

$$Ca(OH)_2 + 2HCl \rightarrow CaCl_2 + 2H_2O \qquad (3.85)$$

$$CaCO_3 + 2HCl \rightarrow CaCl_2 + H_2O + CO_2 \qquad (3.86)$$

$$Ca(OH)_2 + 2HF \rightarrow CaF_2 + 2H_2O \qquad (3.87)$$

$$CaCO_3 + 2HF \rightarrow CaF_2 + H_2O + CO_2 \qquad (3.88)$$

for which a lower temperature is needed than for reactions (3.82) and (3.83) to obtain a maximum rate.

Washing of flue gases by water solution of ammonia can remove SO_2 according to reactions:

$$NH_3 + SO_2 + H_2O \rightarrow NH_4HSO_3 \qquad (3.89)$$

$$NH_4HSO_3 + NH_3 \rightarrow (NH_4)_2SO_3 \qquad (3.90)$$

$$(NH_4)_2SO_3 + \frac{1}{2}O_2 \rightarrow (NH_4)_2SO_4 \qquad (3.91)$$

and the side reactions removing halogens:

$$NH_3 + HCl \rightarrow NH_4Cl \qquad (3.92)$$

$$NH_3 + HF \rightarrow NH_4F \qquad (3.93)$$

Among possible methods of NO_x removal, the conversion of NO_x by NH_3 injected to the combustion gases is the most efficient. At temperatures 920–1000°C with some excess of NH_3 it leads to high conversion. If the temperature of combustion gases is too high, then NO_x is formed rather than destroyed, while for too low temperature the ammonia does not react and is emitted. Despite the presence of large quantities of O_2, ammonia reacts with NO according to reaction

$$NH_3 + NO + 0.25\,O_2 \rightarrow 1.5\,H_2O + N_2 \qquad (3.94)$$

The excess NH_3 is oxidized to nitrogen in the competing reaction

$$4NH_3 + 3O_2 \rightarrow 3N_2 + 6H_2O \qquad (3.95)$$

Breen and Sotter (1978) quote NO_x reduction in a commercial boiler by NH_3 injection equal 30% at $NH_3/NO_x = 1$ and 70% at $NH_3/NO_x = 4$. They report also a possible ammonia utilization of 65% during experiments in demonstration boiler in which NO_x concentration was reduced from 226 to 78 ppm. In order to reduce the consumption of NH_3 catalyzers are proposed. A selective catalyzer containing titanum oxide (SCR) can be applied at temperature 350–450°C and a nonselective active coke at temperatures 80–130°C.

Conversion of NO in presence of SCR catalyzer can proceed according to reaction (3.94) and parallelly in small extend a catalytic reaction

$$SO_2 + \frac{1}{2}O_2 \rightarrow SO_3 \qquad (3.96)$$

Figure 3.33 presents the influence of catalizator temperature on NO_x conversion. At 0.5% oxygen it is possible to obtain NO_x conversion over 80%. The highest efficiency of SCR catalyzer is within the region 350–450°C.

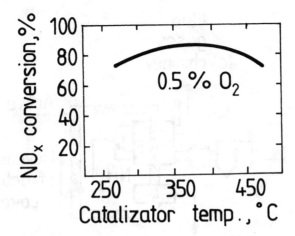

Fig. 3.33. Conversion of NO_x as a function of SCR catalizator temperature for $NH_3/NO_x = 1$ (Jüntgen and Richter, 1985).

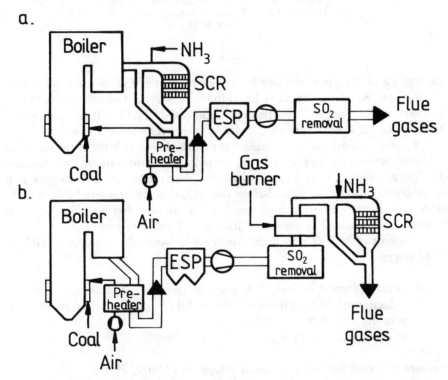

Fig. 3.34. Two ways of integration of the NO_x selective reduction installation with the boiler (Jüntgen and Richter, 1985).

The processes of desulphurization were developed earlier than the processes of NO_x removal, so in many boiler units the NO_x conversion installations are built beyond the desulphurization equipment. Figure 3.34 presents two ways of integration of the NO_x unit with the boiler. In solution "a" the air preheater is beyond the NO_x reduction so then the corrosive salt NH_4HSO_4, which can be created as a product of reaction between excess

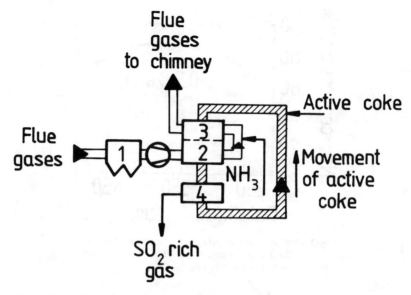

Fig. 3.35. Parallel removal of SO_2 and NO_x by active coke: 1, electrostatic precipitator; 2, removal of SO_2; 3, catalitic reaction of SO_2; 4, regeneration of active coke (Jüntgen and Richter, 1985).

ammonia and SO_3, may deposit in the air preheater and in the electrostatic precipitator. What is more, at reduced load the flye gas temperature is so low that even within the catalyzer the deposition of NH_4HSO_4 salt can take place. In such cases it is necessary to use the by-pass presented it the figure.

These problems can be avoided if the SCR of NO_x is built at the so called cool-end, beyond the air preheater, as presented in bottom figure "b". Because the flue gas temperature beyond the desulphurization unit is too low it is necessary to heat up the gases before the NO_x unit. In presented solution the preheat is done by external fuel, which is an obvious disadvantage, which is however compensated by increased life time of the catalyzer.

The cost of SCR is very high. Applying however the removal of NO_x in three stages:

- direct reduction of NO_x formation in combustion chamber,
- reduction of NO_x to further 50% by NH_3 without catalizator in temperature region 900–1000°C,
- SCR reduction of NO_x,

a reasonable cost has been obtained (Heyn and Harig, 1987).

Active coke can be utilized as adsorber of SO_2 and a catalyzer for NO_x reduction in temperatures 80–130°C (Jüntgen and Richter, 1985). Figure 3.35 presents an installation of parallel adsorption of SO_2 and catalytic reduction of NO_x. Flue gases of temperature 120°C flow through a moving bed of active coke. In the first part of the bed 90% of SO_2 can be removed. Ammonia is injected into the partially cleaned gases, which in the second stage react catalytically with NO_x. At the same time the rest of the SO_2 is adsorbed. Active coke is regenerated in two stages. In the first stage coke is cooled also indirectly and is recycled to the process.

3.7. PULVERIZED COAL BURNERS

A properly designed burner is a unique device guaranteeing effective combustion of coal and a long life time of the combustion chamber. In modern, high capacity combustion chambers single burners are not installed, but groups of multi-nozzle burners. Up to sixty-four burners can be mounted mostly in more than one wind boxes. In proximate considerations one multinozzle burner utilizes 6–10 t/h of coal. To insure good burn-out of pulverized coal the particles must be smaller than 200μm. The smaller is the content of volatiles in coal the finer must be the coal particles. The quality of coal preparation is checked by means of mesh size 200μm and 88μm, which depending on the fuel should produce the following remains on 88μm mesh net:

anthracite	—	(4–6)%,
bituminous coal	—	(30–35)%,
lignite	—	(45–50)%.

The pulverized coal is transported to the combustion chamber by means of flue gases, but most frequently it is done by the primary air. Large amounts of primary air used as transport media cause difficulties with ignition of the primary air-coal mixture, so then for low volatile coals the primary air should be limited to 15% and for high volatile coals up to 25% of the total air.

Coals containing high moisture must be dried, because moisture in coal reduces the flame temperature and can cause unstable combustion. There are two possible solutions of coal drying presented in figure 3.36. Solution "a" can be applied for low moisture coals, usually bituminous coals for which the stechiometric air equals 5–10 kg air/kg coal. For high moisture lignites, however, for which the stechiometric air is equal 3–5 kg air/kg coal, it is necessary to use additional high temperature gases to dry the coal. In practice flue gases of temperature about 1000°C are used, as presented in figure 3.36. The suction of flue gases is forced by a ventilator mill. It is worth noting also that this solution improves also the safety, because the primary air-lignite mixture is easily explosive.

The fuel input to the combustion chamber is limited by the heat flux that can be transferred to the cooling tubes. Typically in a large boiler about half the heat generated during combustion goes mainly by radiation into the water cooled walls of the combustion chamber. One of the important design problems is how densely the burners must be placed. It is not recommended to install the burners at very small distances from each other, because it can cause large temperature gradients within the chamber. The following dimensions should be taken as the minimal values for burners placed on the side walls:

burner axis to side wall	→ 1.5 diameter of the secondary air nozzle,
burner axis to top of bevel funnel	→ 1.5 diameter of the secondary air nozzle,
burner axis to burner axis (horizontal)	→ 3 diameter of the secondary air nozzle,
burner axis to burner axis (vertical)	→ 2.5 diameter of the secondary air nozzle.

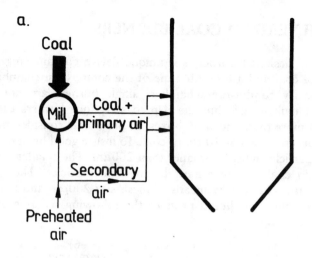

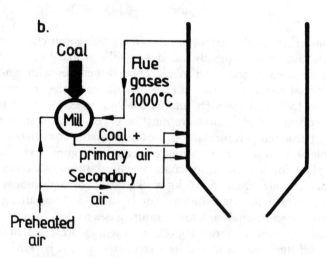

Fig. 3.36. Possible solutions of coal drying:
a, low moisture coal; b, high moisture lignites.

Increased burner spacing influences favourably the NO_x reduction due to more intensive entrainment of cool gases and reduction of temperature gradients. During design of new combustion chambers it is necessary to remember that high intensity of combustion and low NO_x emission are contradictory demands.

Generally pulverized coal burners are divided into two groups:

- swirling jet burners,
- straight jet burners.

The swirling jet burners are recommended for high volatile coals, above 40%(waf). Straight jet burners are applied for low volatile fuels, below 20%(waf). For intermediate volatile content 20–40%(waf) both types of burners can be successfuly applied.

3.7.1. Swirling jet burners

In most practical cases it is not possible to guarantee a constant type of coal during the life time of a boiler. The geometry of the burner must then allow for certain adjustment of the burner to varying coal quality, which usually is done by movable swirling elements. In proximate design the following parameters can be assumed:

fuel stream per one burner	—(6–7) t/h,
temperature of primary air-coal mixture	—<100°C,
temperature of secondary air	—(300–315)°C,
primary air velocity at outflow	—(25–27)m/s,
secondary air velocity at outflow	—(30–35)m/s,
concentration of fuel in primary air	—(300–320)g/m^3.

The developed industrial burners differ mainly by the way of air swirling. Examples of four possible solutions are presented in figure 3.37. In case "a" the changes of air swirling can be obtained by shifting the swirling element along the burner axis. Burner "b" is a modification of the former with swirling of air by tangentially adjustable blades. In solution "c" swirling of the secondary air is realized by tangentially adjustable blades and additionally the primary air-coal mixture is directed towards the swirling air by means of a special element in the centre of the burner. In burner "d" both streams are swirled so that the primary air-coal swirling number is constant and the secondary air swirling can be controlled by adjustable screw-type blades.

For swirling burners the minimal for proper ignition internal recirculation degree, defined as the ratio of recirculating gas stream to the stream of gas flowing out of the burner, is equal to 0.05. Figures 3.38 and 3.39 demonstrate that for burners "c" and "d" the demanded proper ignition conditions and recirculation degree can be obtained. It is worth noting that in the case of burner "d" this can be reached at an angle of the secondary air blades equal to $\Theta_2 = 50–60$ degree, which limits the available range of control. For burner "c" the 0.05 value of the internal recirculation degree can be obtained already at very small swirling number.

In order to increase substantially the size of the central recirculation zone a divergent quarl should be fitted to a burner mouth. The length of the quarl is equal to about 1/2 of the outflow diameter and the half-angle depending on the swirling number:

$$S < 0.5 \rightarrow 20–25 \text{ degrees},$$
$$S > 0.5 \rightarrow 20–35 \text{ degrees}.$$

3.7.2. Straight jet burners

Most of the nonswirling burners are tangentially installed in modern combustion chambers. In order to insure good mixing of coal with air, multinozzle burners were developed. The primary air-coal nozzles can have circular or rectangular cross sections, and usually they are cooled by a "coating" air. The burner has a shape of a vertical column of height (5–7m) above and

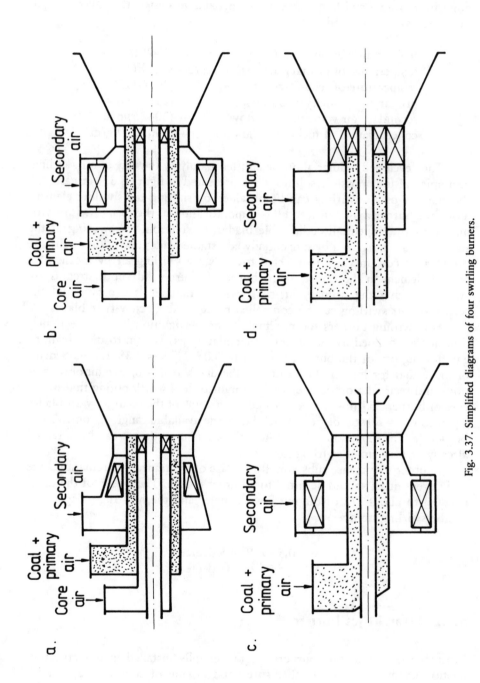

Fig. 3.37. Simplified diagrams of four swirling burners.

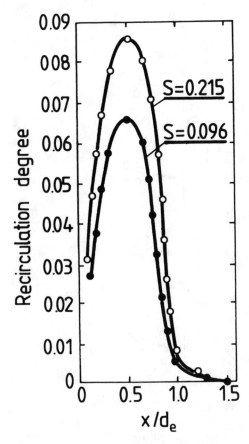

Fig. 3.38. Degree of inner recirculation for burner *c* (Wróblewska et al., 1970).

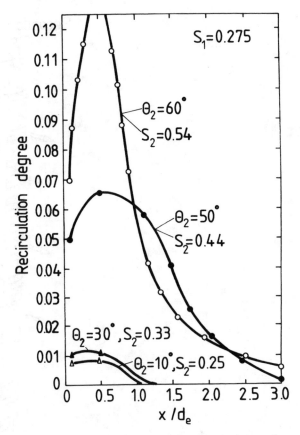

Fig. 3.39. Degree of inner recirculation for burner *d*: S_2, Θ_2—secondary air swirling number and angle (Wróblewska et al., 1970).

below which the nozzles of secondary air are placed. Figure 3.40 presents an example of a burner configuration. The most common parameters of straight jet burners for proximate calculations can be assumed as:

- fuel stream per one burner unit—(6–10)t/h,
- temperature of primary air-coal mixture—<100°C,
- temperature of secondary air—(300–315)°C,
- primary air velocity at outflow—(25–30)m/s,
- secondary air velocity at outflow—(40–50)m/s,
- tilting (downwards) of primary air-coal nozzles—(10–20) degree,
- tilting (downwards) of the upper secondary air nozzles—up to 50 degree,
- tilting of lower secondary air nozzles—(0 ∓ 10) degree.

In some solutions burners are built in the form of double vertical columns between which an intermediate nozzle of secondary air is placed. The most typical division of air can be assumed as:

- upper secondary air nozzle—25%,
- intermediate secondary air nozzle—15%,
- bottom secondary air nozzle—20%,

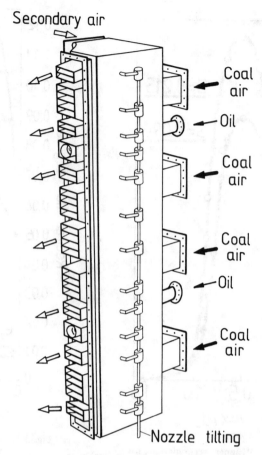

Fig. 3.40. Straight jet burner.

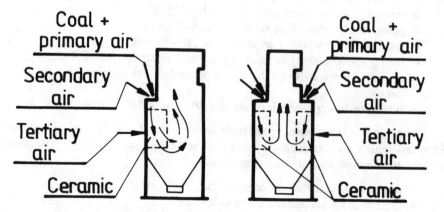

Fig. 3.41. Straight jet burners producing *U* and *W* flames (Cutress and Pierce, 1965).

- coating secondary air nozzles—rest, 20%,
- primary air nozzles—20%.

In case of fuels containing very small volatiles, such as anthracite or coke, the straight jet burners have been successfully used. For such fuels roof firing combustion chambers, presented in figure 3.41, producing U-shape or W-shape flames, were positively tested for pulverized anthracite with 15% moisture and 32% ash (Cutress and Pierce, 1965). The jet burner is equipped with a primary air-coal nozzle and a secondary air nozzle. Along the flame tertiary air is supplied through the nozzles on the side wall of the chamber. To guarantee a proper stability of the flame at (25–100)% of the nominal load, a part of the chamber wall in the neighbourhood of the flame is not cooled. Experiments have shown that up to 40% of the total wall surface area of the combustion chamber must be not cooled to obtain proper stabilization. In case of W-shape flame the tertiary air can be removed, because the burning zone created in the centre of the chamber can stabilize the flames through radiation. The burner roof is usually placed on the level of a half of the total height of the combustion chamber (Cutress and Pierce, 1965).

REFERENCES TO CHAPTER 3

Beér J. H. and Chigier N. A.—Combustion aerodynamics. Applied Science Publishers, London, 1974.

Beér J. H., Chigier N. A. and Lee K. B.—Proc. 9th Symp. (Int).on Combustion, Academic Press, New York, 1963.

Boow J. and Goard P. R.—J. Inst. Fuel, 42(1969)346.

Bowman C. T.—Progr. Energ. Comb. Sci., 1(1975)33–45.

Breen B. P. and Sotter J. G.—Prog. Energy Comb. Sci., 4(1978)201–220.

Chedaille J., Leuckel W. and Chesters A. K.—J. Inst. Fuel, 39(1966)311, 506–521.

Chigier N. A. and Chervinsky A.—Trans. ASME, Ser. E. J. Appl. Mech., 34(1967)443–451.

Chugtai M. Y. and Michelfelder S.—BWK, 35, Marz(1983)75–83.

Craya A. and Curtet R.—Compt. Rend. A. S., Paris, 241,1(1955)621–622.

Cutress J. O. and Pierce T. J.—J. Inst. Fuel, 38, February (1965)54.

Fenimore C. P.—Comb. and Flame, 25 (1975)85.

Field M. A., Gill D. W., Morgan B. B. and Hawksley P. G. W.—Combustion of Pulverized Coal, BCURA, Leatherhead, 1967.

Flament G.—Direct SO$_2$ capture in flames through the injection of sorbent. Doc. F09/a/24, April, 1981, IFRF.

Gibson M. M. and Morgan B. B.—J. Inst. Fuel, 43(1970)517–523.

Gosman A.D., Lockwood F. C. and Salooja A. P.—Proc. 17th Symp. (Int.) on Combustion. The Combustion Institute, Pittsburgh, 1978.

Gosman A.D., Pun W. M., Runchal A. K., Spalding D. B. and Wolfshtein M.—Heat and mass transfer in recirculating flows. Academic Press, London, 1969.

Hannes K.—BWK, Bd.41, 7/8 (1989)328–331.

Hedley A. B. and Jackson E. W.—J. Inst. Fuel, 38(1965)290.

Hedley A. B. and Jackson E. W.—J. Inst. Fuel, 39(1966)208–218.

Heyn K. and Harig H. D.—Entschwefelung und Entstickung von Rauchgasen aus Steinkohlenfeuerungen in der BRD. Steinkohlentag, 9.12.1987, Essen.

Hzmalian D. M. and Kagan Ya.A.—Teoria Gorenia i Topochnie Ustroistva. Energia, Moscow, 1976.

Johnson G. M., Smith M. Y. and Mulcahy M. F. R.—Proc. 17th (Int.) Symp. on Combustion. The Comb. Inst., Pittsburgh, 1979.

Jüntgen H. and Richter E.—BWK, Dokumentation Rauchgasreinigung, Sept. (1985)8–20.

Lavoie G. A., Heywood J. B. and Keck J. C.—Comb. Sci. Technology, 1(1970)313.

Lockwood F. C., Salooja A. P. and Syed S. A.—Combustion and Flame, 38(1980)1–15.

Mc Kenzie A., Smith I. W. and Donau Szpindler G.—J. Inst. Fuel, 47(1974))75–81.

Padia S., Sarofim A. F. and Howard J. B.—The behaviour of ash in pulverized coal under simulated combustion conditions. Paper presented at Combustion Institute's Central States Section Spring Meeting, 5–7 April (1977).

Reid W. T. in Elliott M. A. (ed.)—Chemistry of Coal Utilization. A. Wiley—Interscience Publication. New York, 1981.

Richter W. and Saha R. K.—Arch. Termod. Spalania, 10(1979)485–512.

Ricou F. P. and Spalding D. B.—Journ. Fluid. Mech., 11(1961)21–23.

Sunavala P. D.—J. Inst. Fuel, 41(1968)477–483.

Sunavala P. D.—J. Inst. Fuel, 46(1973)30–38.

Spalding D. B.—VDI Forschungsheft, 549(1972)5–16.

Taylor T. E.—Journ. of Engineering for Power. October(1976)528–539.

Thring H. W. and Newby M. P.—Proc. 4th Symp. (Int.) on Combustion. Williams and Wilkins, Baltimore, 1953.

Tsai S. C.—Fundamentals of Coal Benefication and Utilization. Coal Science and Technology. Elsevier. N. York, 1982.

Ulrich G. D., Riehl J. W., French B. R. and Desrosiers R.—The mechanism of submicron fly ash formation in a cyclone coal fired boiler. Int. Conf. Ash Deposits and Corrosion from Impurities in Combustion Gases. New England College (ASME Research Committee), June 26–July 1, 1977.

Van der Hegge Zijnen B. B.—Appl. Sci. Res., A.1(1958)435–461.

Wróblewska V., Borzyma E. and Dąbrowski K.—Report Inst. Energ., Nr 8345, Warsaw, 1970.

Zeldovich Ja.B., Sadownikow P.Ja. and Frank-Kamenetskij D. A.—Oxidation of nitrogen during combustion. Akademia Nauk USSR. Inst. Chem. Phys., Moscow-1947.

4
COMBUSTION OF COAL
IN FLUIDIZED BEDS

Intensive heat and mass transfer within the fluidized beds create attractive possibilities for coal combustion. The main advantages of this way of combustion can be summarized as follows:

- possible combustion of high ash coals,
- high volumetric loads of combustion chambers,
- easy desulphurization of combustion gases within the combustion chamber,
- reduced nitrogen oxides emission,
- possible combustion of broad size range coal,
- removal of ash in the form of not sintered particles.

Most of the above advantages are caused directly by the low temperature of combustion within the fluidized bed. This temperature is limited from above by the sintering temperature of coal ash and from below by the degree of coal burn-out. Theoretically these limits are within 700–1100°C; nevertheless practically the most suitable temperature range is 850–950°C.

It is important to point out that the ash content within the bed differs very much from that of the feed coal. The fluidized bed within a combustor is formed mainly of ash or in some particular cases of inert particles. The mass content of carbon within the bed is at average on the level of about 1–2%.

The first fluidized bed technologies were based on stationary beds in which some 50% of heat released during combustion was removed by cooled tubes submerged within the fluidizing bed. The main disadvantage of this solution is a large number of necessary feeding points to insure even distribution of coal within the bed, limiting the power of combustion units. This has been overcome by circulating fluidized beds applying much higher velocity of fluidization. Removal of tubes from the bed to avoid metal errosion at high velocity intensified the coal mixing and did allow a reduction of the number of feeding points. Both the above mentioned ways of fluidized bed combustion have been developed to pressurized technologies creating new possibilities of application in gaseous-steam power stations.

4.1. GASODYNAMICS OF FLUIDIZED BEDS

A fluid flowing upwards through a stagnant layer of particles at certain velocity can cause loosening of the particles which can start to move. This phenomenon is called fluidization and the fluid velocity at which it begins is named a minimum fluidization velocity.

If the fluidizing solid particles are evenly distributed within the bed, then we have a regular fluidization. However, gaseous fluidization most frequently is irregular due to gaseous bubbles within the bed. In case the bubbles occupy the whole bed cross-section, they can split the bed into parts moving upwards and gradually dissipating. This kind of irregularity is called slugging. Slugs

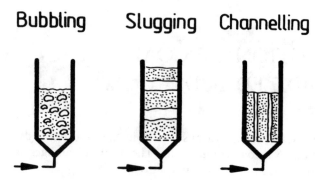

Fig. 4.1. Irregularities of fluidization.

of dense phase, separated by gaseous bubbles, move upwards. Slugging is associated with intensive pulsation of the bed height and pressure at the bottom of the bed. Slugging takes place mainly in beds with high ratios of the bed height to the bed diameter. Another kind of irregularity, channelling, can happen mainly for very small particles. Despite high velocity the stream of gas is unable to initiate fluidization, but flows instead through channels formed within the solid bed. Figure 4.1 presents the three typical kinds of irregular fluidization.

Increasing the fluidization velocity causes bed expansion, so long as the velocity does not exceed the solid particle free fall velocity. Once the velocity exceeds the terminal velocity, the process of fluidization changes into pneumatic transport and in consequence the particles are blown out. There is a common convention to determine the fluidization velocity as an apparent velocity, that means on the basis of an empty reactor.

4.1.1. Minimum fluidization velocity

The flow of fluid through a stagnant bed is associated with the pressure drop of the fluid. According to Kozeny (1927) and Carman (1956) the drop of fluid pressure caused by particles of diameter d_p and shape factor φ, forming bed of height $h_{B,o}$, can be calculated by equation

$$\Delta p = \lambda_f \rho_g \frac{w^2}{2} \frac{h_{B,o}}{d_p} f(\varepsilon_o, \varphi) \qquad (4.1)$$

where the function $f(\varepsilon, \varphi)$ is determined experimentally. Leva et al. (1948) presented equation

$$f(\varepsilon, \varphi) = \frac{(1 - \varepsilon)^{3-n}}{\varepsilon^3} \varphi^{3-n} \qquad (4.2)$$

and a graphical function for the friction number λ_f as well as for the exponent n, shown in figure 4.2. For small Reynolds number $\left(\mathrm{Re} = \dfrac{w d_p}{\nu_g} < 10 \right)$ the flow of fluid is laminar and

$$\lambda_f = \frac{400}{\mathrm{Re}} \qquad (4.3)$$

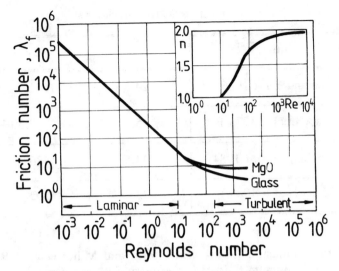

Fig. 4.2. Friction number as a function of Reynolds number.

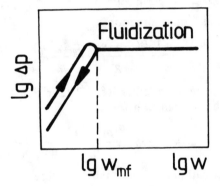

Fig. 4.3. Bed pressure drop as a function
of fluidization velocity.

while for Re > 1000 the flow is turbulent and the friction number is equal

$$\lambda_f = \frac{7}{Re^{0.1}} \tag{4.4}$$

In case of rough particle surface the friction number depends on the roughness and can be twice higher than the values calculated by equations (4.3) and (4.4).

Increasing the fluid velocity we can initiate fluidization of particles and at moderate intensity of fluidization the hydrostatic pressure of the bed is equal

$$\Delta p = h_B(1 - \varepsilon)(\rho_p - \rho_g)g \tag{4.5}$$

Experiments have shown that the change from stagnant to fluidizing bed is associated with a characterisitic hump, presented in figure 4.3. Decrease of fluidization velocity does not show this hump. That means it is formed by denser placement of particles within a stagnant bed than in a fluidizing bed

Table 4.1 **Constants in Ergun equation (4.8).**

Authors	C_1	C_2
Wen and Yu (1966)	24.5	1652
Babu et al. (1978)	15.42	778.50
Thonglimp et al. (1984)	23.5	1486
Zeng et al. (1985)	32	1200

brought into a standstill. Ciborowski (1948) and Miller et al. (1951) have shown that the minimum fluidization velocity does not depend on the mass of the bed and its height. At the point of change from a stagnant bed into a fluidized bed the pressure drop of fluid flowing through a bed must be equal to the hydrostatic pressure of a fluidized bed of height $h_{B,mf}$. This equality is known in literature as Ergun equation

$$\frac{1.75}{\varphi \varepsilon_{mf}^3} \left(\frac{d_p w_{mf} \rho_g}{\eta_g} \right)^2 + \frac{150(1 - \varepsilon_{mf})}{\varepsilon_{mf}^2 \varphi^2} \frac{d_p w_{mf} \rho_g}{\eta_g} = \frac{d_p^3 (\rho_p - \rho_g) \rho_g g}{\eta_g^2} \qquad (4.6)$$

in which for small Reynolds number Re $<$ 20 the first term is negligible, and equation (4.6) can be written in a form

$$w_{mf} = 0.005 \frac{d_p^2 (\rho_p - \rho_g) g}{\eta_g} \frac{\varphi^2 \varepsilon_{mf}^3}{1 - \varepsilon_{mf}} \qquad (4.7)$$

The general Ergun equation (4.6) is usually presented in a dimensionless form

$$C_1 \text{Re}_{mf}^2 + C_2 \text{Re}_{mf} = \text{Ar} \qquad (4.8)$$

where the Reynolds number is for the minimum fluidization velocity and the Archimedes number is equal

$$\text{Ar} = \frac{g \, d_p^3 (\rho_p - \rho_g)}{\nu_g^2 \rho_g} \qquad (4.9)$$

A number of developed constants C_1 and C_2 are presented in table 4.1.

In case of beds containing particles of broad size, n size groups of diameter $d_{p,i}$ representing group i-th with mass fraction g_i, Leva recommends the equivalent particle diameter

$$d_p = \left(\sum_{i=1}^{n} g_i / d_{p,i} \right)^{-1} \qquad (4.10)$$

The most frequently quoted set of constants in equation (4.8) are those by Wen and Yu (1966) which are recommended for all fluids and solid particles with shape factor $\varphi = 1.0$–6.7. In case of small Reynolds number ($\text{Re}_{mf} < 20$) the minimum fluidization velocity is then equal

$$w_{mf} = \frac{d_p^2 (\rho_p - \rho_g) g}{1650 \, \eta_g} \qquad (4.11)$$

Table 4.2 **Shape Factors.**

Particle	φ	Source
Coal, $0 \div 6$ mm	1.37	Podkowa (1970)
Pulverized Coal	1.45	Shirai (1954)
Anthracite, Bituminous Coal	1.6	Leva et al. (1948)
Coal, $0.5 \div 6$ mm	1.69	Kudzia (1980)
Ash from fluidized bed, <2.5 mm	3.3	Lorkiewicz and Jastrząb (1982)
Sand	3.4	Brotz (1964)

The Babu et al. (1978) constants are recommended for fluidization of carbonaseous particles (coal, char, coke).

For nonspherical particles the effective particle diameter is defined as a diameter of a sphere with the volume equal to the particle volume V_p

$$d_p = \sqrt[3]{\frac{6}{\pi} V_p} \qquad (4.12)$$

The divergence of the particle surface area A_p from the surface area of a sphere of diameter equal to the effective diameter (eq.4.12), determines the shape factor

$$\varphi = \frac{A_p}{\pi d_p^2} \qquad (4.13)$$

Table 4.2. presents the shape factors of various particles.

4.1.2. Terminal fluidization velocity

Increasing the value of fluidization velocity we reach a point at which the solid particles are blown out by pneumatic transport. At this point the fluid velocity is equal to the free fall velocity of the solid particles in a stagnant fluid, and the velocity is called a terminal fluidization velocity w_t. The porosity of the bed at velocity w_t is assumed to be equal to unity.

A solid particle characterized by a drag number λ_d will reach a free fall velocity which can be calculated by a balance of forces: drag force, gravita-

Table 4.3 **Drag Number in Intermediate Region (Ciborowski, 1957).**

Shape Factor	Reynolds Number				
	1	10	100	400	1000
1.50	28	6.0	2.2	2.0	2.0
1.25	27	5.0	1.3	1.0	1.1
1.18	27	4.5	1.2	0.9	1.0
1.06	27	4.5	1.1	0.8	0.8

tion, and buoyancy, which has a form

$$\lambda_d \rho_g \frac{w_t^2}{2} \frac{\pi d_p^2}{4} = \frac{\pi d_p^2}{6} (\rho_g - \rho_p)g \qquad (4.14)$$

For spherical particles the drag number is a function of the Reynolds number. In case of small Reynolds numbers Re < 0.4, and even for Re < 2, the flow of fluid around the falling particle is laminar and

$$\lambda_d = \frac{24}{Re} \qquad (4.15)$$

For intermediate Reynolds number (2–500) the drag number is equal

$$\lambda_d = \frac{18.5}{Re^{0.6}} \qquad (4.16)$$

while for Re > 500 the flow of fluid is turbulent and the drag number is constant $\lambda_d = 0.43$.

In case of nonspherical particles the value of λ_d depends on both the Reynolds number and the particle shape factor. Ciborowski (1957) proposed for a laminar fall

$$\lambda_d = \frac{28.5}{\lg \left(\dfrac{15.4}{\varphi} \right) Re} \qquad (4.17)$$

and for a turbulent fall

$$\lambda_d = 5.31 - \frac{4.88}{\varphi} \qquad (4.18)$$

For the intermediate region Ciborowski recommends the value of λ_d from table 4.3.

Applying a proper equation describing the drag number, it is possible to calculate the terminal fluidization velocity by means of equation (4.14). An analytical formula can be obtained for laminar (Re < 0.4) and turbulent (Re > 500) flow of fluid around the falling particle. For the intermediate region (0.4 < Re < 500) Kunii and Levenspiel (1969) simplified the equation describing the drag number for spherical particles into a form

$$\lambda_d = \frac{10}{\text{Re}^{0.5}} \tag{4.19}$$

and obtained equation

$$w_t = \left(\frac{4}{225} \frac{(\rho_p - \rho_g)^2 d_p^3 g^2}{\rho_g \, \eta_g} \right)^{1/3} \tag{4.20}$$

4.1.3. Porosity of fluidized beds

The porosity of a bed is defined as a ratio of the free volume to the total volume of the bed. A stagnant bed porosity is slightly smaller than at minimum fluidization velocity. Knowing the density of solid particles ρ_p, the density of fluidizing gas ρ_g, and the density of the bed immediately after fluidization termination ρ_t, we can calculate the porosity of a bed at minimum fluidization velocity

$$\varepsilon_{mf} = \frac{\rho_p - \rho_t}{\rho_p - \rho_g} \tag{4.21}$$

For spherical particles the value of ε_{mf} is a function of the particle placement within the bed. Beds of monosized particles in regular placements have the following values of ε_{mf} (Bennet and Myers, 1967):

$$
\begin{aligned}
\text{tetrahedrite} \quad &- \quad \varepsilon_{mf} = 0.26, \\
\text{rhombic} \quad &- \quad \varepsilon_{mf} = 0.3, \\
\text{ortorhombic} \quad &- \quad \varepsilon_{mf} = 0.4, \\
\text{cubic} \quad &- \quad \varepsilon_{mf} = 0.48.
\end{aligned}
$$

Experiments with spherical monosized particles show the value $\varepsilon_{mf} = 0.4$ which indicates that the ortorhombic placement is the most favourable. For small spherical particles (<0.3mm) the value of ε_{mf} increases but does not exceed 0.48.

The porosity of a bed at minimum fluidization velocity depends on the ratio of particle diameter. For dry particles and for coals with moisture content below ambient air equilibrium moisture, we can assume that, in case of particle diameter to bed diameter ratio smaller than 0.1, the porosity is equal to that in an infinitely large bed. Kudzia (1980) developed an experimental equation describing the relation between the value of ε_{mf} and the coal particle diameter

$$\varepsilon_{mf} = \frac{0.140}{1 + d_p} + 0.462 \tag{4.22}$$

where d_p should be used in mm.

The porosity of a fluidized bed increases with velocity, from ε_{mf} to unity. According to Lewis et al. (1949) for a regular fluidization it is possible to describe this relation by equation

$$\frac{w}{w_{mf}} = \varepsilon^n \tag{4.23}$$

Coal with moisture smaller than the analytical moisture follows this experimental relation. For a 18% moisture in coal a regular fluidization can be obtained at porosity higher than 0.8.

One of the fundamental problems is the influence of temperature and pressure on porosity at minimum fluidization velocity. Wen et al. (1985) and Pattipati and Wen (1982) found a very weak dependence of ε_{mf} on pressure and temperature and suggest assuming it to be equal to that at room parameters.

4.1.4. Segregation of particles in fluidized beds

There are two possible kinds of nonhomogenuity of particles within the bed:

- particles of the same density but of different size,
- particles of the same size but of different density.

Shannon (1959) noticed that segregation of particles of different density but of equal diameter depends mainly on the value of minimum fluidization velocity of particle groups and on the gas velocity. He proposed a ratio of the w_{mf} values of two successive groups of particles to be smaller than 2, as a criterion for maintaining the bed in fluidization state.

Rowe et al. (1972) reported that segregation of particles is sensitive mainly to density differences and less to diameter differences. The segregation has been shown to be proportional to the diameter ratio (larger to smaller) to the power of 0.5 and to the density ratio (larger to smaller) to power of 2.5.

Chen and Keairns (1975) examined a mixture of dolomite (2800 kg/m³) and coke (702 kg/m³) within particle size 0.18–2.8mm. Applying large fluidization numbers $\dfrac{w}{w_{mf}} = 1.5-2$ of the larger or denser particles, they have obtained very good mixing for beds of height equal to the bed diameter. Reduction of the gas velocity to values close to the minimum fluidization velocity of larger or denser particles initiated segregation. Larger or denser particles concentrated on the bottom, while the smaller or lighter particles concentrated on the top. Increase of bed height to an extent causing slugging favoured segregation.

The rate of particle segregation within the fluidized bed is high, already after 30s a semi-steady state particle distribution can be obtained. The increase of pressure to 0.6–1.0 MPa does reduce the tendency to separation because the w_{mf} value of larger particles decreases with pressure, while with smaller particles this velocity is almost independent of pressure.

4.1.5. Gaseous bubbles

Gaseous fluidization is generally heterogeneous because a large part of the fluidizing gas flows through the bed in the form of bubbles. The movement of bubbles and their growth have decisive influence on heat and mass transfer mechanisms within the fluidizing bed. It is possible to distinguish along the bed height regions of bubble formation, growth, and splitting.

A single isolated gaseous bubble will rise in a stagnant unlimited liquid

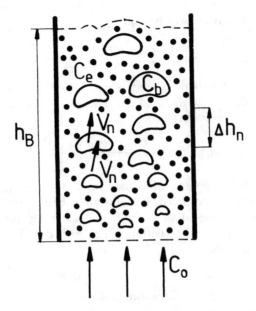

Fig. 4.4. Rise of gaseous bubbles in a fluidized bed.

with velocity

$$w_b = 0.71(g\, d_b)^{1/2} \tag{4.24}$$

In case of bubbles rising in a fluidized bed various equations were proposed for the absolute velocity of bubbles, among which the equation by Davidson and Harrison (1963) is most often quoted

$$w_a = w - w_{mf} + w_b \tag{4.25}$$

Also a simplified relation is quoted

$$w_a = 1.2\, w_b \tag{4.26}$$

There are some ways of calculation of bubble size in a fluidized bed, among which the theoretical model by Darton (1979) is readily used. This model utilizes the two phase theory of fluidization assuming that beds are formed of two parts:

- emulsion phase,
- bubbles.

The total stream of the fluidizing gas is distributed among these two parts so that through emulsion phase gas flows with minimum fluidization velocity and the rest flows through the bed in form of bubbles.

Darton assumed that bubbles rise in fluidized bed in certain "paths" formed over the distributor ports presented in figure 4.4. If the cross-section area of bed per one path of bubbles is equal A_{B1}, then the volume stream flowing within one path of bubbles is equal

$$\dot V = (w - w_{mf})A_{B1} \tag{4.27}$$

If the vertical distance between the bubbles is equal to their frontal diameter

d_{bf}, then the volume stream will be equal

$$\dot{V} = \frac{V_b w_b}{d_{bf}} \tag{4.28}$$

Because the volume of a single bubble is equal $V_b = \pi d_b^3/6$ and for a hemispherical bubble $d_{bf} = 1.26 \, d_b$, then for a single bubble velocity according to equation (4.24) Darton obtained

$$d_b = 1.63 \left[\frac{(w - w_{mf}) A_{B1}}{g^{1/2}} \right]^{2/5} \tag{4.29}$$

The theory of bubbles rising assumes that bubbles of certain size live within a distance proportional to horizontal distance between the neighbouring streams of bubbles. The diameter of the catching up area is equal $(4 \, A_{B1}/\pi)^{1/2}$ and so the distance within which n-th collision takes place can be calculated by equation

$$\Delta h_n = 0.72(w - w_{mf})^{-1/2}(g \, d_{bn})^{1/4} \tag{4.30}$$

where d_{bn} is a bubble diameter after n-th collision and 0.72 is an experimental constant.

The height of n-th collision over the distributor is equal

$$h_n = 0.72(w - w_{mf})^{-1/2} g^{1/4} \sum_{i=1}^{n-1} d_{bi}^{5/4} \tag{4.31}$$

where the sum can be replaced by a continuous integral.

If a gas volume is not lost during collision of bubbles, then $d_{bn} = 2 \, d_{b(n-1)}$. Assuming next that the diameter of the first bubble can be calculated from equation (4.29), replacing in it the value A_{B1} by the surface area A_o of a one distributor port, Darton obtained the bubble diameter at height h

$$d_b = 0.54(w - w_{mf})^{2/5} \frac{(h + 4\sqrt{A_o})^{4/5}}{g^{1/5}} \tag{4.32}$$

Mori and Wen (1975) published relation allowing us to calculate the bubble diameter d_b on the height h above the distributor

$$\frac{d_b - d_{bm}}{d_{bo} - d_{bm}} = \exp\left(-0.3 \frac{h - h_{bo}}{d_B} \right) \tag{4.33}$$

where the initial and maximum bubble diameters within a bed of diameter d_B are equal

$$d_{bo} = 3.77(w - w_{mf})^2/g \tag{4.34}$$

$$d_{bm} = 2.59 \, g^{-0.2}[(w - w_{mf}) A_B]^{0.4} \tag{4.35}$$

The initial height h_{bo}, on which the bubbles of diameter d_{bo} are formed, is equal to the jet height at the outflow from the distributor (eq.4.123).

The bed expansion due to the presence of bubbles is determined by the difference $h_B - h_{B,mf}$. Darton utilizing equations (4.31) and (4.32) obtained equation

$$h_B - h_{B,mf} = 0.6 \, g^{-1/4}(w - w_{mf})^{1/2} \sum_{i=0}^{n-1} d_{bi}^{3/4} \tag{4.36}$$

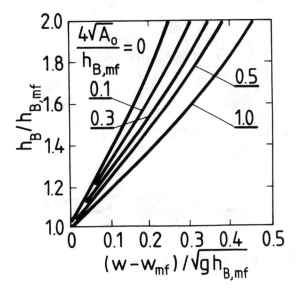

Fig. 4.5. Expansion of a bubbling bed (Darton, 1979).

which can be presented in a form

$$\frac{h_B}{h_{B,mf}} = 1 + 2\frac{(w - w_{mf})^{4/5}}{(g\,h_{B,mf})^{2/5}}\left[\left(\frac{4\sqrt{A_o}}{h_{B,mf}} + \frac{h_B}{h_{B,mf}}\right)^{3/5} - \left(\frac{4\sqrt{A_o}}{h_{B,mf}}\right)^{3/5}\right] \quad (4.37)$$

Figure 4.5 presents the bed expansion calculated by means of the last equation.

4.2. HEAT AND MASS TRANSFER IN FLUIDIZED BEDS

Heat and mass are transferred within a fluidized bed combustor due to temperature and concentration differences:

particle—emulsion gas—bubbles.

Mass transfer coefficient β between spherical particle and gas can be calculated by means of the Sherwood number given in a form

$$\mathrm{Sh} = \frac{\beta\,d_p}{D} = 2 + a\,\mathrm{Re}^b\mathrm{Sc}^c \quad (4.38)$$

In fluidized bed combustors the burning ash particle is surrounded by emulsion phase formed of ash particles as presented in figure 4.6. The presence of ash particles slows down the mass diffusion so that the effective diffusion coefficient within emulsion is smaller than the molecular diffusion coefficient D. Jung and La Nauze (1983) published an equation for Sherwood number allowing us to calculate the mass transfer coefficient between the particle and emulsion gas

$$\mathrm{Sh} = \frac{\beta\,d_p}{D} = 2\,\varepsilon + 0.69\left(\frac{\mathrm{Re}}{\varepsilon}\right)^{0.5}(\mathrm{Sc})^{1/3} \quad (4.39)$$

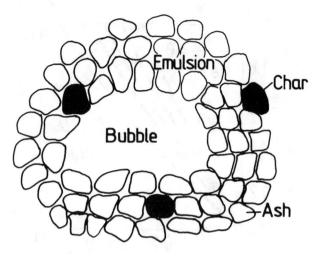

Fig. 4.6. Ash particles diluted in emulsion phase.

4.2.1. Mass transfer between emulsion and bubbles

Bubbles exchange substance with gas in emulsion phase. If the volume stream from the bubbles to emulsion is equal $\dot{V}$ then a similar stream flows from emulsion to bubbles in order to fulfil the two phase theory.

Davidson and Harrison (1963) came to the conclusion that the stream of gas flowing between the bubble and emulsion is formed of convection and diffusion parts

$$\dot{V} = \pi d_b^2 \left[w_{mf} + 0.76 \frac{\varepsilon_{mf}}{1 + \varepsilon_{mf}} D^{1/2} \left(\frac{g}{d_b} \right)^{1/4} \right] \qquad (4.40)$$

The variation of gas concentration within the bubble can be calculated from the gas balance equation. Assuming ideal mixing of emulsion phase we can write

$$\dot{V}(C_b - C_e) = -w_a V_b \frac{dC_b}{dh} \qquad (4.41)$$

It is possible to integrate the last equation within a distance of i-th collision where V_{bi} and w_{ai} are constant. The gas concentration in a bubble flowing out of the bed to the freeboard can then be calculated by equation

$$\frac{C_b(h_B) - C_e}{C_o - C_e} = \exp \left(-\sum_{i=1}^{N} \frac{\dot{V}_i \Delta h_i}{w_{ai} V_{bi}} \right) \qquad (4.42)$$

where C_o is the gas concentration at the outflow from the distrtibutor and N the number of collision distances along the bed height. The sum on the right hand side of the last equation (4.42) denotes an index of transversal flow, meaning the number of times the gas volume within the bubble is renewed during rising from the bottom to bed height h_B.

Introducing a mass transfer coefficient β_{be} defined by equation

$$\dot{V}(C_b - C_e) = \pi d_b^2 \beta_{be}(C_b - C_e) \qquad (4.43)$$

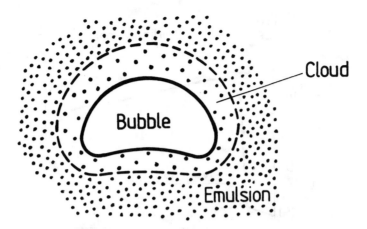

Fig. 4.7. Gaseous bubble surrounded by a cloud.

and utilizing equation (4.40) we can write that the mass transfer coefficient between the bubble and emulsion is equal

$$\beta_{be} = w_{mf} + 0.76 \frac{\varepsilon_{mf}}{1 + \varepsilon_{mf}} D^{1/2} \left(\frac{g}{d_b}\right)^{1/4} \tag{4.44}$$

Some investigators claim that the above equation gives too high values of mass transfer coefficient.

Kunii and Levenspiel (1969) assumed presence of a third body called "cloud" between the bubble and emulsion (Fig. 4.7). In this model the mass transfer coefficient between the bubble and the cloud is also described by convection and diffusion terms in a form

$$\beta_{bc} = 0.75 \, w_{mf} + 0.975 \, D^{1/2} \left(\frac{g}{d_b}\right)^{1/4} \tag{4.45}$$

The coefficient between the cloud and the surrounding emulsion is given by a diffusion term

$$\beta_{ce} = 2 \left(\frac{D \varepsilon_{mf} w_b}{\pi d_b}\right)^{1/2} \tag{4.46}$$

The mass transfer coefficient between the bubble and emulsion can be then calculated

$$\frac{1}{\beta_{be}} = \frac{1}{\beta_{bc}} + \frac{1}{\beta_{ce}} \tag{4.47}$$

The above equations were developed on the basis of experiments with single bubbles. Sit and Grace (1981) presented an equation developed on the basis of experiments in a two-dimensional bed

$$\beta_{be} = \frac{w_{mf}}{3} + 2 \left(\frac{D \varepsilon_{mf} w_b}{\pi d_b}\right)^{1/2} \tag{4.48}$$

Very often a value $\beta_{be} = 0.016$ m/s is used for light particles ($\rho_p \leqslant 1000$ kg/m^3) and $\beta_{be} = 0.009$ m/s for heavy particles ($\rho_p > 1000$ kg/m^3) (Kobayashi et al., 1967).

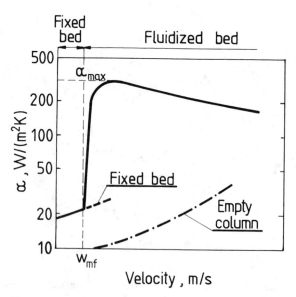

Fig. 4.8. Variation of heat transfer coefficient with fluidization velocity.

4.2.2. Heat transfer between fluidized bed and immersed surface

Mixing of solid particles within the fluidized beds causes increase of the heat transfer coefficient between the bed and the surfaces of heat exchangers. Figure 4.8 presents an example of a typical variation of heat transfer coefficient with fluidization velocity. For a fixed bed the coefficient is higher then for an empty column flow and appears as a monotonic increasing function with gas velocity. Beyond the minimum fluidization velocity the heat transfer coefficient increases steeply, reaching a minimum and decreasing monotonicaly beyond it for higher velocities. Karchenko and Makhorin (1964) reported the maximum value of heat transfer coefficient at gas velocity $(2-5)w_{mf}$. However, for beds with particles in broad size range, that always takes place during combustion, even up to $8\ w_{mf}$ (Mc Lahren and Williams, 1969) and $9\ w_{mf}$ (Petrie et al., 1968) the maximum was not obtained.

The value of maximum heat transfer coefficient is a monotonic function of solid particle diameter as presented in figure 4.9. This relation has been published first by Baerg et al. (1950) and confirmed later in many experiments. For very small particles the maximum value of heat transfer coefficient can be as high as 800 $W/(m^2K)$. However, for (2–3)mm particles the maximum value for atmospheric pressure and room temperature is about 200 $W/(m^2K)$. According to Botterill (1975) the maximum value of heat transfer coefficient can be taken to be proportional to a mean power (-0.35) of the particle diameter, reaching a minimum value at particle diameter (2–3)mm. For larger particles the coefficient tends to increase again (Zabrodsky, 1966).

The maximum heat transfer coefficient depends on the physical properties of gas and on the volumetric heat capacity of the solid particles, but they seem to be independent of the thermal conductivity of the solid (Wicke and Fatting, 1954).

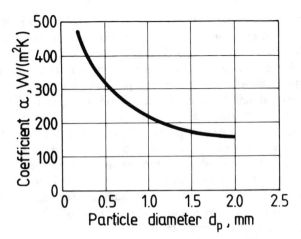

Fig. 4.9. Heat transfer coefficient as a function of particle diameter.

Many theoretical models of heat transfer within fluidized beds were developed, among which the model by Mickley and Fairbanks (1955) is most often quoted. However the existing models describe qualitatively the influence of various physical values on heat transfer, it is still necessary to use experimental equations for designing purposes.

Heat is transferred between the solid surface and the fluidized bed by mechanisms which can be regarded as essentially independent of each other: particle convection, gas convection, and radiation. The heat transfer coefficient is then formed of three parts:

$$\alpha = \alpha_p + \alpha_g + \alpha_r \qquad (4.49)$$

The term "particle convection" refers to the mechanism of energy transfer through the moving particles. Experiments have shown that the majority of heat is transferred by conduction through a thin gaseous gap between the particles and the surface and only a very small part by conduction in contact points with the surface. The size of the gaseous gap is from $\frac{1}{10} d_p$ (Botterill and Williams, 1963) to $\frac{1}{6} d_p$ (Zabrodsky, 1966). Conduction of heat through the gaseous gap is not a limiting factor for heat transfer, but the residence time of particles in the surface neighbourhood is a limiting factor. Due to small heat capacity of particles they cool (or heat) quickly to the temperature of the surface, slowing down the rate of heat transfer.

Gas convection is the direct gas to surface heat transfer by these parts of the surface which are not in contact with the particles. The gas convection heat transfer coefficient increases with increasing gas velocity. Baskakov et al. (1973) developed an experimental correlation

$$Nu_g = \frac{\alpha_g d_p}{\lambda_g} = 0.009 \, Pr^{1/3} \, Ar^{1/2} \left(\frac{w}{w_{opt}}\right)^N \qquad (4.50)$$

where w_{opt} is the fluidization velocity corresponding to the maximum value of the total heat transfer coefficient. For velocity $w \geq w_{opt}$ the power factor should be taken $N = 0$, while for $w < w_{opt}$ a value $N = 0.3$ is recommended. The gas convection component becomes dominant at high fluidization velocities, that takes place for large particles.

Combustion of coal in fluidized beds is carried out at temperatures 800–1000°C, so the radiation mechanism can play an important role. The contribution of radiation has been a subject of much research. Botterill (1970), analysing results of various authors, found the contribution of radiation equals (5–10)% at bed temperature 500°C and (50–60)% at bed temperature 1400°C. Later experiments by Szekely and Fischer (1969) have shown that the contribution of radiation is important at temperatures above 1000°C. Yoshida et al. (1974) confirmed that below 1000°C bed temperature the contribution of radiation is very small. They assumed also that the linear increase of heat transfer coefficient with temperature (below 1000°C) is caused by the increase of gas thermal conductivity.

The radiation heat transfer coefficient between bed of temperature T_e and wall of temperature T_w can be calculated by equation

$$\alpha_r = \varepsilon_w \, \sigma \, \frac{T_e^4 - T_w^4}{T_e - T_w} \approx 4 \, \varepsilon_w \sigma T_m^3 \qquad (4.51)$$

where $T_m = \dfrac{T_w + T_e}{2}$ is the medium bed-wall temperature. Equation (4.51) is based on the assumption that the whole surface area exchanges heat with the fluidizing bed. This is only partially true, because the particles close to the surface have temperature different from the bed temperature T_e. Consequently then, the radiation heat transfer coefficient calculated by means of equation (4.51) should be taken as the maximum possible value.

A fundamental contribution to particle convection heat transfer modelling has been made by Mickley and Fairbanks (1955). Their model, known commonly as a "packet model," assumes that the bulk of solid particles in a fluidized bed is completely mixed, so that the bulk temperature T_e is uniform within the bed. Packets of emulsion, containing particles at minimum fluidization state, are driven through the bed by rising bubbles. Once the packet approaches a surface of temperature $T_w < T_e$, a nonsteady temperature field forms within the packet, as presented in figure 4.10. The temperature of particles close to the surface decreases. The rate of temperature variation depends on the particle diameter and its thermal capacity. The final temperature the particles can reach during the contact with the surface is a function of both the rate of temperature variation and the contact time duration t_C. The mobile packets of emulsion residing at the surface during time t_C move back into the bulk of the bed. It is believed that renewal of the surface by fresh packets is a result of bubbles rising.

Determination of the packet-surface contact time is not an easy task. Most frequently the contact time is related to the surface linear dimension (vertical) and to the fluidization velocity

$$t_C = \frac{L}{w} \qquad (4.52)$$

or

$$t_C = \frac{L}{w - w_{mf}} \qquad (4.53)$$

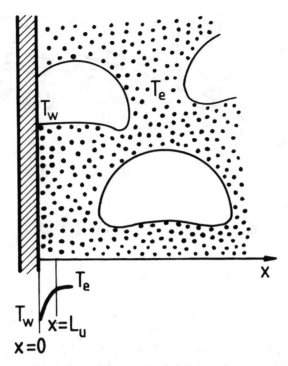

Fig. 4.10. Temperature field within a packet of emulsion
contacting the solid surface.

but such a parameter is rather a measure of gas contact time than the particle
contact time. Because the movement of packets in bed is initiated by bubbles
rising with frequency n_b, then the contact time of particles with the heat
transfer surface can be calculated by equation (Yoshida et al., 1969)

$$t_C = \frac{1 - \varepsilon_b}{n_b} \qquad (4.54)$$

where ε_b is the volume fraction of bubbles in bed.

The packet model (Mickley and Fairbanks, 1955) predicts a maximum
value of heat transfer at the instant contact time. Yoshida et al. (1969) presented a solution of the unsteady temperature field within a packet, shown
in figure 4.10. If the heat conduction coefficient through the emulsion is equal
to λ_e and its heat capacity $(\rho c)_e$ then the heat transfer equation is

$$\frac{\partial T}{\partial t} = \frac{\lambda_e}{(\rho c)_e} \frac{\partial^2 T}{\partial x^2} \qquad (4.55)$$

The initial conditions: $t = 0$, $T(x,0) = T_e$ and the boundary conditions:
$x = 0$, $T(0,t) = T_w$; $x = L_u$, $T(L_u,t) = T_e$.

For short contact times for which

$$\frac{1}{L_u} \left(\frac{\lambda_e}{(\rho c)_e} t_C \right)^{1/2} < 1 \qquad (4.56)$$

Yoshida et al. obtained a solution of equation (4.55) allowing them to determine the particle heat transfer coefficient, which was proposed originally

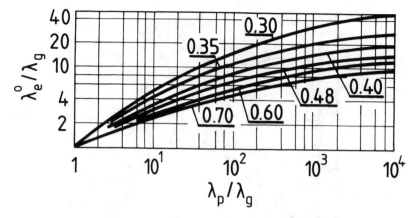

Fig. 4.11. Thermal conductivity in a stagnant bed (Baskakov, 1966).

by Mickley and Fairbanks

$$\alpha_p = \left(\frac{\lambda_e(\rho c)_e}{\pi t_C}\right)^{1/2} \tag{4.57}$$

The emulsion parameters are equal: $L_u \approx (1-3)d_p$, $(\rho c)_e \approx (\rho c)_p(1 - \varepsilon_{mf})$. Thermal conductivity of emulsion phase can be calculated by Yagi and Kunii (1967) equation

$$\lambda_e = \lambda_e^o + 0.1(\rho c)_g \, d_p w_{mf} \tag{4.58}$$

where for $\lambda_p/\lambda_g \leqslant 5000$ the thermal conductivity of the stagnant packed bed is related to the gas thermal conductivity (Gelpierin et al., 1966)

$$\lambda_e^o = \lambda_g \left[1 + \frac{(1 - \varepsilon)\left(1 - \frac{\lambda_g}{\lambda_p}\right)}{\frac{\lambda_g}{\lambda_p} + 0.28 \, \varepsilon^{0.63} \left(\frac{\lambda_p}{\lambda_g}\right)^{0.18}}\right] \tag{4.59}$$

or in general case can be found in figure 4.11 (Baskakov, 1966).

Martin (1984) modified the Mickley and Fairbanks (1955) model by application of a molecular rather than "continuum" concept. He calculated the particle convection heat transfer coefficient α_p to be dependent on the mean velocity of random displacement of particles between bed bulk and the surface.

For practical applications many empirical correlations were developed, among which that developed by Zabrodsky (1966) is very often quoted

$$\alpha_{max} = 35.8 \, \lambda_g^{0.6} \rho_p^{0.2} \, d_p^{-0.36} \tag{4.60}$$

in which if all values are used in principal SI dimensions we get the heat transfer coefficient in $W/(m^2K)$. Equation (4.60) has been tested positively by many authors, but a general reservation concerning this equation arises from the dimensional analysis. In most cases such equations should be used with caution. What is more, the Zabrodsky equation indicates that the value of α is independent of heat exchanger tube diameter. Experiments show, however,

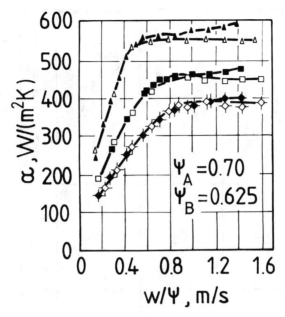

Fig. 4.12. Heat transfer coefficient between bed and horizontal tubes as a function of pressure and velocity for 5 heating elements (Bock and Schweinzer, 1985):

	A	B
1MPa	◆	◈
1.5MPa	■	□
2.5MPa	▲	△

that for tube diameter 60mm the heat transfer coefficient is larger than for tubes 35mm (Mc Lahren and Williams, 1969).

Gas pressure has an effect on heat transfer coefficient, mainly for large particle diameters. This influence is caused mainly by gas convection contribution due to density depending on pressure. At higher pressures, as long as $\rho_g \ll \rho_p$, it is possible to find by means of equation (4.50) a square root relation

$$\frac{\alpha_g(p)}{\alpha_g(p_o)} = \left(\frac{p}{p_o}\right)^{1/2} \tag{4.61}$$

For higher pressures the gas convection becomes more important in the heat transfer mechanism. Bock and Schweinzer (1985) measured heat transfer coefficient between a square cross-section bed (0.4 × 0.4m) and a bundle of 42.2mm horizontal tubes at three pressures: 1.0, 1.5 and 2.5 MPa. Two configurations of the heat exchanger were used: A-20 tubes, horizontal distance—110.5mm and vertical distance—95.7mm; B-32 tubes, horizontal distance—88.4 mm and vertical distance—76.6mm. Results for 5.3mm particles are presented in figure 4.12. The value ψ means a degree of cross-section decrease due to the presence of tubes.

Distance between the tubes is an important design parameter. For a single tube the heat transfer coefficient has a maximum value. If the distance between the tube axis is smaller than two diameters, the value of α decreases

Table 4.4 Experimental Nusselt Number Correlations.

Geometry	Equation	Authors	Remarks
Horizontal Tube	$Nu_{max} = 0.21 \cdot Ar^{0.32}$	Baskakov et al. (1973)	monodiameter particles, characteristic diameter - d_p
Horizontal Tube	$Nu = 47 \cdot (1-\varepsilon) \cdot \left(\dfrac{Re}{Ar}\right)^{0.325} \cdot Pr^{0.30} \cdot \left(\dfrac{\varrho_p \cdot c_p \cdot d_t^{\frac{3}{2}}}{\lambda_g} \cdot g^{0.5}\right)^{0.23}$	Grewal and Saxena (1980)	$Nu = \dfrac{\alpha d_t}{\lambda_g}$, $Re = \dfrac{w d_t}{\nu_g}$
	$Nu_{max} = 0.79 \cdot Ar^{0.22} \cdot \left(1 - \dfrac{d_t}{s_1}\right)^{0.25}$	Gelpierin et al. (1968b)	$s_1/d_t = 2 \div 9$ characteristic diameter - d_p
	$Nu_{max} = 0.74 \cdot Ar^{0.22} \cdot \left[1 - \dfrac{d_t}{s_1}\left(1 + \dfrac{d_t}{s_2 + d_t}\right)\right]^{0.25}$	Gelpierin et al. (1968a)	$s_1/d_t = 2 \div 9$ $s_2/d_t = 0 \div 10$ characteristic diameter - d_p
	$Nu_{max} = 0.9 \cdot \left(Ar \cdot \dfrac{d_{t_0}}{d_t}\right)^{0.21} \cdot \left(\dfrac{c_p}{c_{pg}}\right)^{0.2} \cdot \left[1 - 0.21\left(\dfrac{s}{d_t}\right)^{-1.75}\right]$	Grewal and Saxena (1983)	$d_{t_0} = 12.7$ mm characteristic diameter - d_p $s/d_t = 1.75 \div 9$ $75 \leq Ar \leq 20000$
Vertical Tube	$Nu_{max} = (0.116 \cdot Ar^{0.3} + 0.0175 \cdot Ar^{0.46}) \cdot Pr^{0.33}$	Borodulya et al. (1980)	characteristic diameter - d_p
Horizontal Tube	$Nu = 0.66 \cdot Pr^{0.3} \cdot \left[Re \cdot \dfrac{\varrho_p}{\varrho_g} \cdot \dfrac{1-\varepsilon}{\varepsilon}\right]^{0.44}$	in Kunii and Levenspiel (1969)	$Nu = \dfrac{\alpha d_t}{\lambda_g}$, $Re = \dfrac{w d_t}{\nu_g}$ $Re < 2000$ characteristic diameter - d_t

Table 4.4 Experimental Nusselt Number Correlations. (Continued)

Geometry	Equation	Authors	Remarks
Horizontal Tube	$Nu = 420 \cdot \left(\dfrac{Re}{Ar}\right)^{0.3} \cdot Pr^{0.3}$	Vreedenberg (1958)	$Nu = \dfrac{\alpha \cdot d_t}{\lambda_g}$, $Re = \dfrac{w \cdot d_t}{v_g}$
Horizontal Tube	$Nu = 900 \cdot (1-\varepsilon) \cdot \left(\dfrac{Re}{Ar}\right)^{0.326} \cdot Pr^{0.3}$	Andeen and Glicksman (1976)	$Nu = \dfrac{\alpha \cdot d_t}{\lambda_g}$, $Re = \dfrac{w \cdot d_t}{v_g}$
Horizontal Tube	$Nu = 14 \cdot \left(\dfrac{w}{w_{mf}}\right)^{\frac{1}{3}} \cdot Pr^{\frac{1}{3}} \cdot \left(\dfrac{d_t}{d_p}\right)^{\frac{2}{3}}$	Petrie et al. (1968)	$Nu = \dfrac{\alpha \cdot d_t}{\lambda_g}$
Horizontal Tube	$Nu = 5.76 \cdot (1-\varepsilon) \cdot \left(\dfrac{Re}{\varepsilon}\right)^{0.34} \cdot Pr^{0.33} \cdot \dfrac{h_{B,0}^{0.16}}{d_B} \cdot \dfrac{d_t}{d_p}$	Ainstein (1966)	$Nu = \dfrac{\alpha \cdot d_t}{\lambda_g}$, $Re = \dfrac{w \cdot d_p}{v_g}$
Horizontal Tube	$Nu = 4.38 \cdot \left[\dfrac{1}{6 \cdot (1-\varepsilon)} \cdot Re\right]^{0.32} \cdot \dfrac{1-\varepsilon}{\varepsilon} \cdot \dfrac{d_t}{d_p}$	Gelperin et al. (1966)	$Nu = \dfrac{\alpha \cdot d_t}{\lambda_g}$, $Re = \dfrac{w \cdot d_p}{v_g}$
Horizontal Tube	$Nu_{max} = 0.9 \cdot \left(Ar \cdot \dfrac{d_{t_0}}{d_t}\right)^{0.21} \cdot \left(\dfrac{c_p}{c_{pg}}\right)^{45.5 \cdot Ar^{-0.7}}$	Grewal and Saxena (1981)	$d_{t_0} = 12.7mm$ characteristic diameter $-d_p$ $75 \leq Ar \leq 20000$
Horizontal Tube	$Nu = 2.9 \cdot \left(\dfrac{1-\varepsilon}{\varepsilon} \cdot Re\right)^{0.4} \cdot Pr^{0.33} \cdot \dfrac{d_t}{d_p}$	Ternovskaya and Korenberg (1971)	$Nu = \dfrac{\alpha \cdot d_t}{\lambda_g}$, $Re = \dfrac{w \cdot d_p}{v_g}$

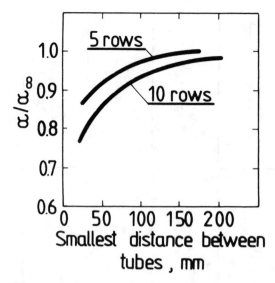

Fig. 4.13. Influence of the distance between the 60
mm tubes on the heat transfer coefficient
(Mc Lahren and Williams, 1969).

sharply. Figure 4.13. presents the ratio of the heat transfer coefficients for a
tube bundle and for a single tube.

Some experimental correlations for heat transfer coefficient are collected
in table 4.4. For reasons discussed before only equations in which the dimensional analysis is correct are included.

4.3. DIFFUSION OF PARTICLES WITHIN
FLUIDIZED BEDS

Feeding coal into fluidized bed combustion creates gradients of fuel concentration. Mixing of solid particles within the bed is very intensive. Nevertheless
for large bed cross-section areas the concentration gradients can be so high
that the released coal volatiles can reach the freeboard before they mix with
air and burn out.

Mixing of particles within fluidized beds is caused mainly by the movement of bubbles. Experiments show that lateral mixing is much slower than
the vertical one (Kunii and Levenspiel, 1969a). Consequently large gradients
of concentration and temperature should be expected in lateral direction.
Immersed tubes increase the diffusional resistance, and this has to be taken
into account to avoid large gradients. It has been shown that the rate of
mixing intensifies with the size of the bed (Emielianov et al., 1967).

Movement of solid particles in direction "x" within the fluidized bed can
be described by means of a diffusion equation

$$\frac{\partial C}{\partial t} = D \frac{\partial^2 C}{\partial x^2} \tag{4.62}$$

This process is determined by random motion of bubbles of known size and
velocity. The lateral diffusion coefficient of solid particles can then be derived

from equations of statistical physics, which allow us to relate it to the fluidization conditions.

4.3.1. Lateral diffusion of particles in a free bed

Borodulya et al. (1982) assumed that bubbles flowing through emulsion phase cause turbulent pulsations of particles and consequently their lateral movement, resulting in value of a diffusion coefficient to be proportional to the mixing path of the "vortex" and the average velocity of the turbulent pulsations of particles. Based on this assumption and experiments they have obtained a relation between the diffusion coefficient and the fluidization conditions

$$D = 0.013 g^{0.15} h_{B,mf}^{0.65} d_B^{0.5} (w - w_{mf})^{0.7} \qquad (4.63)$$

Some investigators interpret lateral displacement of solid particles, caused by the motion of bubbles, in terms of the Einstein random walk theory. According to this theory the diffusion coefficient for a one-dimensional movement may be expressed by the mean-square displacement $\overline{(\Delta x)^2}$ during time interval Δt in a form

$$D = \frac{1}{2} \frac{\overline{(\Delta x)^2}}{\Delta t} \qquad (4.64)$$

Based on this approach Kunii and Levenspiel (1969b) obtained the lateral (radial) diffusion coefficient

$$D = \frac{3}{16} \left(\frac{\varepsilon_b}{1 - \varepsilon_b} \right) \frac{w_{mf} d_b}{\varepsilon_{mf}} \qquad (4.65)$$

The approach of Kunii and Levenspiel has been used also by Bellgardt (1985) for large beds.

Shi and Fan (1985) assumed that the bubbles moving through the emulsion phase press on the particles, causing their lateral displacement. These random walk displacements are the reason for lateral diffusion of particles. Applying equation (4.64) they obtained a formula

$$D = 0.0625 d_b w_b \qquad (4.66)$$

which can be transformed into relation

$$\frac{(w - w_{mf}) h_{B,mf}}{D} = 22.5 \left[\frac{w - w_{mf}}{(g h_{B,mf})^{0.5}} \right]^{0.25} \qquad (4.67)$$

There are also available equations for diffusion coefficient obtained through statistical correlation of experimental data. Shi and Fan (1985) on the basis of investigations in a shallow bed obtained equation

$$\frac{D}{(w - w_{mf}) h_{B,mf}} = 0.46 \left[\frac{(w - w_{mf}) d_p \rho_g}{\eta_g} \right]^{-0.21} \left[\frac{h_{B,mf}}{d_p} \right]^{0.24} \left[\frac{\rho_p - \rho_g}{\rho_g} \right]^{-0.43} \qquad (4.68)$$

Chmielewski and Selecki (1977) on the basis of experiments in a small rectangular bed height, presented a relation

$$D = 10^{-3} h_{B,o} w_{mf} \left(\frac{w}{w_{mf}} \right)^{4.5} \qquad (4.69)$$

Table 4.5 **Solid Paticles Diffusion Coefficient in a Pilot Scale
Fluidized Bed** (Highley and Merrick, 1971).

Fluidization Velocity, m/s	0.61				1.07			
Tubes in Bed	Yes		No		Yes		No	
Bed Height, m	0.305	0.610	0.305	0.610	0.305	0.610	0.305	0.610
Diffusion Coefficient, $D \cdot 10^4$, m²/s	39.0	47.4	74.4	79.1	48.4	79.1	111.6	148.8

4.3.2. Lateral diffusion of particles in a bed with immersed tubes

The presence of tubes within the bed decreases the rate of solid particles mixing. Merrick and Highley (1971) measured the diffusion coefficient in a pilot-scale fluidized bed of 1.5 diameter. The results for beds with and without the immersed heat exchanger are presented in table 4.5. For beds of height 0.61m with two rows of tubes the upper 0.15m of the bed is without tubes. The measured values of diffusion coefficient show the following relations:

- within the velocity range 0.6–1 m/s the diffusion coefficient is proportional to fluidization velocity,
- for beds both with and without an immersed heat exchanger the increase of bed height causes an increase of the diffusion coefficient.

Tomeczek et al. (1991), for vertical tube bundles forming a symmetrical triangular pitch with distance s between the rows of tubes, developed a formula on the basis of equation (4.64)

$$D = \frac{\pi - 2\gamma - \sin 2\gamma}{16(\pi - 2\gamma)}(2s - d_t)w_b \tag{4.70}$$

where

$$\gamma = \arcsin\left(\frac{d_t}{2s}\right) \tag{4.71}$$

They have observed that the bubbles velocity is close to that calculated for slugging regime (Davidson and Harrison, 1971)

$$w_b = w - w_{mf} + 0.35[g(2s - d_t)]^{0.5} \tag{4.72}$$

in which the bubble diameter is assumed $(2s - d_t)$. Figure 4.14 presents experimental values of diffusion coefficient for a free bed and for a bed with a vertical tube bundle. The immersed tubes $\frac{2s}{d_t} \approx 2.5$ drastically reduce the dif-

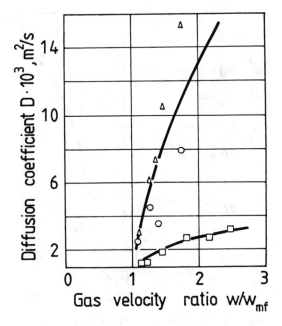

Fig. 4.14. Lateral diffusion coefficient as a function of

fluidization number $\dfrac{w}{w_{mf}}$ for:

— free bed, $\triangle$—coal, $\circ$—char,
— bed with triangular pitch of vertical

tubes, $\dfrac{2s}{d_t} \approx 2.5$, $\square$—coal.

fusion coefficient; what is more its dependence on velocity is very weak in contrast to that in a free bed.

4.3.3. Coal concentration profile

A uniform distribution of coal and air within the bed is a condition of good burn-out and combustion without local over heating. For practical reasons coal is fed to the combustor in a limited number of points and distributed throughout the bed by diffusion. The profile of coal concentration can be calculated by means of local coal concentration balance equation

$$\frac{dC}{dt} = \frac{dC}{dt}\bigg|_{\text{diffusion}} + \frac{dC}{dt}\bigg|_{\text{combustion}} + \frac{dC}{dt}\bigg|_{\text{elutriation}} \qquad (4.73)$$

in which the three terms can be taken to be independent. Highley and Merrick (1971) solved the above equation and the results are presented in figure 4.15. We can see clearly that for a distance over 1m between the feeding points, the concentration of coal at the neighbourhood of feeding is so high that oxygen is completely consumed. Consequently then, along the majority of bed height a reducing atmosphere exists leading to emission of carbon oxide and even methane. In contrast, far from the feeding points oxidizing conditions exist along the whole bed height. The material of tubes in fluidized beds must then resist both the oxidizing and reducing atmospheres.

The optimum number of feeding points must be found by economic

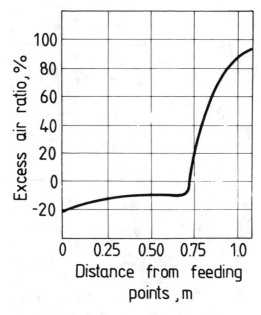

Fig. 4.15. Calculated values of lateral local air excess
ratio (Highley and Merrick, 1971).

analysis; however, in proximate calculations one feeding point per not more
than 2m² of air distributor surface area can be assumed.

4.4. ENTRAINMENT AND ATTRITION OF PARTICLES

The solid particle size distribution within the bed differs from that of the
feeding coal. During the design of a new fluidized bed combustor the following important parameters have to be determined:

- particle size distribution in bed,
- size distribution of elutriated particles,
- elutriation rate.

It is important to notice that the chemical composition of solids in a
fluidized bed combustor is drastically different from that of coal. Simplifying,
the bed of the combustor is built mainly of inert particles (ash) in contrast
to coal where the mineral matter does not exceed 30%. To understand it, the
two main independent phenomena causing the particle size variation must be
analyzed:

- combustion of coal particles,
- attrition of inert particles.

The combustion time of average size coal particles in fluidized beds (5–
8mm) is measured in minutes, while the residence time of ash particles is
measured in hours. The cause of the difference is that carbon content in beds
of combustors is at the level of one percent. Observation of bed particles
shape indicates that the dominant mechanism forming the ash particles is
attrition. The following guidelines can be formulated:

Table 4.6 **Solid Size Particle Distribution During Fluidized Bed Combustion of Newstead Coal** (Merrick and Highley, 1972).

Fluidization Velocity Bed Height	Residence Time, h	Sample	Mass Content within the Size Range in μm						
			>2000	2000 ÷1000	1000 ÷500	500 ÷250	250 ÷125	125 ÷63	<63
0.6 m/s 0.6 m	39	Overflow, 17%	0	10	31	46	13	0	0
		Elutriation, 83%	0	0	0	3	28	18	51
		Feeding	0	10	24	21	13	10	22
1.2 m/s 0.6 m	14	Overflow, 10%	4	16	41	39	0	0	0
		Elutriation, 90%	0	0	0	15	18	16	51
		Feeding	6	11	15	14	14	13	27
1.8 m/s 0.75 m	6	Bed	8	28	56	8	0	0	0
		Elutriation, 100%	0	0	4	17	16	14	49
		Feeding	6	11	15	14	14	13	27

- attrition causes mass loss mainly in the form of fine dust,
- there is a large size difference of bed particles and of elutriated particles,
- particles of size twice as big as the terminal size (free fall) for the fluidization velocity are elutriated to cyclones, even for very high freeboards (height equal two bed diameters).

Table 4.6 presents particle size analysis published by Merrick and Highley (1972) for a 0.6m² combustor fed by a mixture of Newstead coal and ash particles. The residence time of solid given in the table is defined as a ratio of the bed equilibrium mass and total ash mass fed into the combustor. A clear influence of velocity on ash residence time can be seen.

4.4.1. Entrainment of particles

A part of the combustor above the bed is commonly called a freeboard. Non-uniform size distribution within the bed causes that particles are lifted up from the bed, and that is known as entrainment. The larger particles entrained to a certain height above the bed surface fall back on it later. The height of entrainment above the bed surface can be related to the high-velocity gas jets produced by the erupting bubbles. Figure 4.16 presents the variation of solids concentration above bed surface. Zeng and Weil (1958) on the basis of jet dissipating velocity proposed the height of particle entrainment to be equal

$$h_e = 18.2 \, d_b \qquad (4.74)$$

Particles are blown out of the freeboard if they are smaller than the size for which the terminal velocity is smaller than the fluidization velocity. Some of the larger particles are also elutriated and do not return to the bed. The concentration of particles above the bed decreases, and it is assumed that at the entrainment height the concentration reaches a constant value.

It is commonly assumed that the rate of mass elutriation of particles of

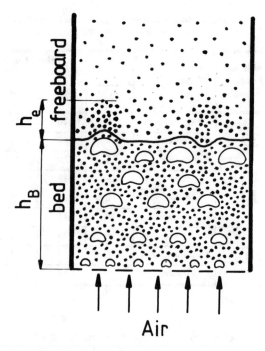

Fig. 4.16. Variation of solids concentration above
the bed surface.

diameter $d_{p,i}$ is proportional to its mass fraction g_i in bed

$$\dot{m}_{ei} = A_B F_i g_i \tag{4.75}$$

The total mass flow of all particles elutriated from the bed surface of cross-section A_B is equal

$$\dot{m}_e = A_B \sum_i F_i g_i \tag{4.76}$$

Yagi and Aochi (1983) proposed an empirical equation for elutriation rate coefficient

$$F_i = \frac{\eta_g (w - w_{ti})^2}{g \, d_{p,i}} (0.0015 \, \text{Re}_{ti}^{0.6} + 0.01 \, \text{Re}_{ti}^{1.2}) \tag{4.77}$$

Merrick and Highley (1972) on the basis of a pilot scale combustor proposed

$$F_i = 130 \, \rho_g w \exp\left[-10.4 \left(\frac{w_{ti}}{w}\right)^{0.5} \left(\frac{w_{mfi}}{w - w_{mfi}}\right)^{0.25} \right] \tag{4.78}$$

The stream of mass flowing upwards decreases with increasing distance above the bed surface. Commonly an exponential function for the rate of entrainment along the freeboard height is assumed

$$\frac{\dot{m} - \dot{m}_e}{\dot{m}_o - \dot{m}_e} = \exp(-ah) \tag{4.79}$$

where $\dot{m}_o$ is the mass stream of solids lifted up from the surface of the bed and a is an empirical coefficient with the average value $a = 4 \, \text{m}^{-1}$ (Large et al., 1976).

Table 4.7 Size Distribution of Fines Generated as Result of Attrition
(Merrick and Highley, 1972).

Particle Size, μm	$0 \div 8$	$8 \div 16$	$16 \div 31$	$31 \div 63$	$63 \div 125$
Mass Content, %	20	20	20	20	20

The mass flow of entrainment at the bed surface can be calculated by
Wen and Chen (1982) equation

$$\dot{m}_o = 3.07 \circ 10^{-9} A_B \dot{m}_g d_b \frac{\rho_g^{2.5} g^{0.5}}{w \eta_g^{2.5}} (w - w_{mf})^{2.5} \tag{4.80}$$

The mass flow rate of particles of diameter $d_{p,i}$ is equal

$$\dot{m}_{oi} = \dot{m}_o g_i \tag{4.81}$$

where g_i is the mass fraction of particles $d_{p,i}$ in bed.

4.4.2. Attrition of particles

Attrition can result from mechanical forces and thermal stresses. It occurs in
two stages, the first instantaneous and the second continuous. Researchers
disagree widely on the effect of parameters on attrition rate (Newby et al.,
1983). Merrick and Highley (1972) developed the following equation allowing the calculation of the mass rate of fines generated as a result of continuous
attrition of fluidized bed material of mass m_B

$$\dot{m}_a = B m_B (w - w_{mf}) \tag{4.82}$$

where the attrition coefficient is equal $B = 9.1 \circ 10^{-6}$ m^{-1} for ash and
$2.7 \circ 10^{-6}$ m^{-1} for limestone.

The experimental evidence disagrees on the dependence of particle size
on attrition. Kono (1979) observed the attrition to be independent of d_p.
Assuming that equation (4.82) can be applied also to any solid particle size
group of medium diameter $d_{p,i}$, we can write

$$\dot{m}_{ai} = B m_{Bi} (w - w_{mf}) \tag{4.83}$$

where m_{Bi} is the mass of group with medium particle diameter $d_{p,i}$ in bed.

Because there are not reliable data allowing us to relate the variation of
attrition coefficient with particle diameter, then assuming it as constant we
can find the rate of particle diameter variation

$$\frac{d(d_{p,i})}{dt} = -\frac{B}{3}(w - w_{mf})d_{p,i} \tag{4.84}$$

Equations (4.82) and (4.83) supply the necessary theoretical background
for particle size modelling with the following assumptions:

- attrition of particles generates fines with size distribution given in table
  4.7,

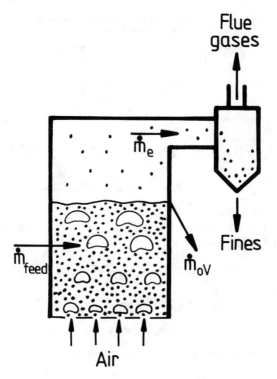

Fig. 4.17. The streams of mass of solids and gases.

- particles of diameter smaller than $d_{pc} = 1.6\, d_{pt}$ are immediately blown out from bed, where d_{pt} is the particle diameter for terminal velocity equal to the fluidization velocity $w_t(d_{pt}) = w$.

Let us consider a fluidized bed combustor in a steady state with the mass streams indicated in figure 4.17. The stream of incombustible solids removed from bed is formed of the following components:

- particles of diameter smaller than d_{pc} at feeding, $\dot{m}_{e1}$;
- fines generated as a result of attrition of coarse particles and immediately elutriated, $\dot{m}_{a1}$;
- particles of diameter smaller than d_{pc} formed as a result of coarse particles attrition $\dot{m}_{a2}$;
- particles removed through overflow, $\dot{m}_{ov}$.

The stream of incombustible fines fed into the combustor and instantaneously elutriated is equal

$$\dot{m}_{e1} = \dot{m}_{feed}\, g_{feed}(d_{pc}) \qquad (4.85)$$

where $g_{feed}(d_{pc})$ is the mass content of incombustible particles smaller than d_{pc} in feed.

The stream of fines generated during attrition is equal

$$\dot{m}_{a1} = B\, m_B (w - w_f) \qquad (4.86)$$

The stream $\dot{m}_{a2}$ elutriated after reduction of particles diameter from d_{po} to d_{pc} can be calculated by means of mass balance for a group of particles.

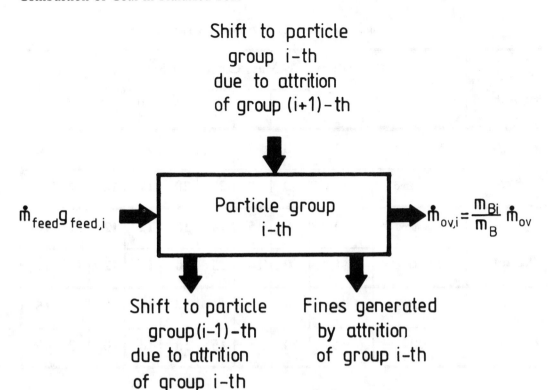

Fig. 4.18. Mass balance for i-th incombustible particle group of $g_{\text{feed}, i}$ mass content at feed.

Let us consider N groups of particles in bed. At steady state conditions the mass of any group in bed is constant, so

$$\frac{dm_{Bi}}{dt} = 0 \qquad (4.87)$$

Writing the mass balances for each group (N equations) according to the diagram presented in figure 4.18, we can find the mass of each group ($i = 1, 2, \ldots N$). The stream of particles elutriated after attrition of initially coarse particles to final diameter d_{pc} is equal

$$\dot{m}_{a2} = \frac{m_{B1}}{t_1} \qquad (4.88)$$

where the attrition time t_1 of particles of group 1 can be calculated by integration of equation (4.84) within the limits of particle diameter of this group.

The total mass flow of elutriated particles is equal

$$\dot{m}_e = \dot{m}_{e1} + \dot{m}_{a1} + \dot{m}_{a2} \qquad (4.89)$$

and the mass stream leaving the bed through the overflow is equal

$$\dot{m}_{ov} = \dot{m}_{\text{feed}} \sum_i g_{\text{feed}, i} - \dot{m}_e \qquad (4.90)$$

The particles within the bed are well mixed, so the size distribution in overflow is equal to that in bed.

Merrick and Highley (1972) calculated the size distribution of elutriated

Table 4.8   **Calculated and Measured Size Distribution of Elutriated Particles**
(Merrick and Highley, 1972).

Velocity Bed Height	Source	Elutriation % of feed ash	Mass Content of Size Group in μm							
			>500	500 ÷250	250 ÷125	125 ÷63	63 ÷32	32 ÷16	16 ÷8	<8
0.6 m/s **0.6 m**	Measured	83	**0**	**3**	**28**	**18**	**12**	**12**	**11**	**16**
	Calculated	84	**0**	**0**	**21**	**20**	**17**	**14**	**14**	**14**
1.2 m/s **0.6 m**	Measured	90	**0**	**15**	**18**	**16**	**12**	**12**	**11**	**16**
	Calculated	91	**0**	**14**	**15**	**20**	**15**	**12**	**12**	**12**
1.8 m/s **0.75 m**	Measured	100	**4**	**17**	**16**	**14**	**11**	**12**	**11**	**15**
	Calculated	100	**10**	**13**	**15**	**18**	**14**	**10**	**10**	**10**

particles and compared it with the measured values in a 0.6m² pilot combustor. The results presented in table 4.8 agree well with the experiments.

4.5. NITROGEN OXIDES EMISSION FROM FLUIDIZED BED COMBUSTORS

Relatively low temperature in fluidized bed combustors should result in small nitrogen oxides emission. However, the emission of NO_x observed in practice can reach even 800 ppm. That is much higher than the equilibrium value for the bed temperature and is close to the level observed for much higher temperatures of pulverized coal flames. The reason for this is the mechanism of NO_x formation:

- oxidation of air-nitrogen,
- oxidation of coal-nitrogen.

The first mechanism is well known and can be described by the classical Zeldovich kinetic model (eqs. (3.71), (3.72)). The second mechanism is much less well understood; however it is commonly accepted that it is much faster than the first and it is responsible for the higher than equilibrium level of NO_x emission. In fluidized bed combustors there is experimental evidence for the second mechanism (Gibbs et al., 1976). At bed temperatures 800–950°C the 90–95% of NO_x forms nitrogen oxide NO and the remaining nitrogen dioxide NO_2 as it leaves the combustor.

Kirner (1945) noticed a release of nitrogen during thermal decomposition of coal in two clear stages. In the first stage of devotalilization about 50% of coal-nitrogen is released with tar volatiles as ammonia and free nitrogen. The remaining 50% of nitrogen is strongly binded in coal and it is not easy

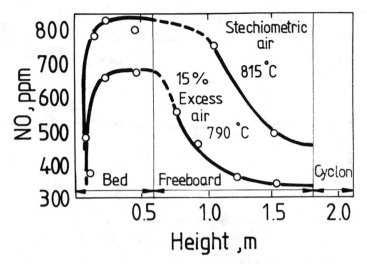

Fig. 4.19. Measured NO concentration along bed height: particle diameter 0.167cm, velocity 0.92m/s (Pereira et al., 1974).

to remove it from the coal structure. There is also no experimental evidence of any bounds of nitrogen with coal ash.

Pereira et al. (1974) measured the profile of NO concentration in a fluidized bed combustor 0.3×0.3 m^2 of height 0.61m, presented in figure 4.19. It can be seen that nitrogen oxide is formed in the lower part of the bed at the main reaction zone where high oxygen concentration exists. The increase of NO concentration with the bed temperature and excess air are expected, but the decrease of NO concentration in the freeboard can only be explained on the basis of reaction with elutriated char. Pereira et al. have found also that at bed temperature 800°C about 10% of nitrogen oxides is formed through oxidation of air nitrogen, and it increases to over 20% at temperature 950°C.

Gradual supply of air to a fluidized bed combustor creates within the bed conditions reducing the nitrogen oxides emission. In this solution the amount of air supplied through the distributor is under-stechiometric and the rest is supplied into the freeboard. Figure 4.20 presents the degree of NO reduction that can be expected. At bed temperature 800°C a 35% reduction can be obtained if 25% of air is supplied into the freeboard as secondary air. This reduction is caused mainly by reaction of NO with carbon, volatiles, and probably with CO, the concentration of which increases in under-stechiometric conditions.

The concentration of nitrogen oxides in flue gases from fluidized bed combustors decreases at elevated pressures. Roberts et al. (1975) reported a reduction of NO$_x$ emission from 300–600 ppm in an atmospheric pressure combustor to 80–130 ppm in a (0.35–0.6)MPa pressurized combustor.

Nitrogen oxides generated during combustion of volatiles and char are subsequently reduced to nitrogen by reactions with:

- char,
- volatile matter,
- nitrogeneous compounds in volatiles,
- carbon monoxide by heterogeneous catalysis.

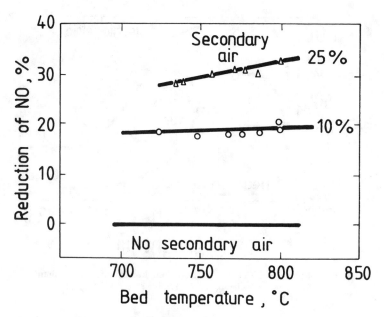

Fig. 4.20. Influence of staged air supply on nitrogen oxides emission (Gibbs et al., 1976).

Nitrogen oxide NO reacts with char to yield CO, CO_2 and N_2 as products. Experiments have shown that at typical fluidized bed temperatures (800–950°C) the main reaction products are CO and N_2 (Gibbs and Hampartsoumian, 1984). Rajan and Wen (1980), however, prefer CO_2 as the product of char and nitrogen oxide reaction

$$C + 2NO \rightarrow CO_2 + N_2 \tag{4.91}$$

with the rate given by equation

$$\dot{R}_{NO} = -A_p \, k_{NO} \, C_{NO} \tag{4.92}$$

where A_p is the surface area of char particles and the rate constant is equal

$$k_{NO} = 5.24 \circ 10^5 \exp(-1.424 \circ 10^8/RT), \text{ m/s} \tag{4.93}$$

Hydrocarbons from volatile matter can play a significant role in NO reduction and hence in its overall emission.

Ammonia released from coal can contribute considerably to NO reduction in fluidized bed combustors. The reaction of ammonia with nitrogen oxides proceeds via a complex free radical chain involving NH_2, NH and N radicals, which can be presented as overall effective reactions:

$$4NH_3 + 6NO \rightarrow 5N_2 + 6H_2O \tag{4.94}$$
$$4NH_3 + 4NO + O_2 \rightarrow 4N_2 + 6H_2O \tag{4.95}$$

The optimum temperature for these reactions is 900–1000°C.

It was found (Shelef and Kummer, 1971) that carbon monoxide can react catalytically with nitrogen oxide on the iron oxide surface according to reaction

$$CO + NO \underset{Fe_2O_3}{\rightarrow} CO_2 + \frac{1}{2}N_2 \tag{4.96}$$

The above reaction can not be excluded in fluidized bed combustors using iron oxide in ash as catalysis.

4.6. SULPHUR BINDING IN FLUIDIZED BED COMBUSTORS

Sulphur oxides formed from both the organic and the mineral matter of coal can be partially reduced by lime from the mineral matter. The amount of sulphur that can be removed from flue gases in this way is, however, too small and additional desulphurization is necessary. This can be done by addition of lime directly into the bed of the combustor where almost ideal temperature conditions exist for sulphur binding by lime.

The addition of lime into a combustor can be done in the form of limestone ($CaCO_3$) or dolomite ($CaCO_3 \circ MgCO_3$) added directly into the bed. Dolomites in general are more reactive than limestones; however the attrition rate of dolomite is higher, so limestone is usually preferred for atmospheric combustors.

Limestone can react directly with SO_2 forming $CaSO_4$ (Locke et al., 1975), but at temperatures of fluidized bed combustors (750°C–850°C) it undergoes calcination according to endothermic reaction

$$CaCO_3 \rightarrow CaO + CO_2 \qquad (4.97)$$

Calcination of limestone proceeds only if the partial pressure of carbon dioxide in combustion gases is smaller than the equilibrium CO_2 pressure of the above reaction at bed temperature. Because the carbon dioxide depends on excess air, the higher is the excess air the lower is the calcination temperature. In atmospheric pressure combustors calcination of limestone takes place at bed temperatures higher than 750°C for excess air ratios applied in combustors.

The calcination of dolomite occurs in stages. During the first step of thermal decomposition at temperatures about 600°C a mixture of calcium and magnesium carbonates is formed

$$CaCO_3 \circ MgCO_3 \rightarrow CaCO_3 + MgCO_3 \qquad (4.98)$$

The magnesium carbonate is then rapidly calcinated already at temperatures 730–760°C

$$CaCO_3 + MgCO_3 \rightarrow CaCO_3 + MgO + CO_2$$

Further calcination of calcium carbonate will take place according to reaction (4.97) appropriate to bed temperature and excess air.

Calcination of limestone or dolomite causes losses of CO_2 and develops porosity of sorbent particles. Rajan and Wen (1980) presented equations describing the development of the specific surface area A_l of limestone during its calcination, correlated with calcination temperature:

$$\text{for } T < 1253 \text{ K}, \qquad A_l = 35.9\,T - 3.67 \circ 10^4, \text{cm}^2/\text{g} \qquad (4.100)$$
$$\text{for } T \geqslant 1253 \text{ K}, \qquad A_l = -38.4\,T + 5.6 \circ 10^4, \text{cm}^2/\text{g} \qquad (4.101)$$

Sulphur dioxide will react in the presence of calcium oxide to form a solid calcium sulphate according to reactions:

$$CaO + SO_2 \rightleftarrows CaSO_3 \tag{4.102}$$

$$CaSO_3 + \frac{1}{2}O_2 \rightleftarrows CaSO_4 \tag{4.103}$$

It has been shown (Bougwardt and Harvey, 1972) that MgO reacts only slowly with SO_2 at temperatures of fluidized bed combustors, so that magnesia oxide can be regarded as an inert substance.

During sulphation the molar volume increases so the sorbent particle loses its porosity. The products of sulphation reaction can block the pores of sorbent particles thus preventing access to the internal surfaces. The pore size distribution is then the main factor determining the sorbent reactivity. Boug-wardt (1970) and Wen and Ishida (1973) propose to calculate the reaction rate of one limestone particle of diameter d_{pl} with sulphur dioxide as

$$\dot{R} = \frac{\pi d_{pl}^3}{6} k\, C_{SO_2} \tag{4.104}$$

The overall reaction rate constant can be calculated by

$$k = k_o \exp(-E/RT)A_l f_l \tag{4.105}$$

where: $k_o = 490$ g/cm^{2}s, $E = 7.33 \circ 10^7$ J/kmol, A_l is the limestone specific surface area and f_l is the limestone conversion factor which takes into account the reduction of limestone reactivity by blocking of pores during sulphation of calcinated limestone, which can be calculated according to Rajan et al. (1978).

The choice of sorbent particle size is a very important factor. Because the reaction products hinder the sorbent particle reactivity dependent on the thickness of the sulphated layer, then the smaller particle size allows for a greater part of the sorbent to be utilized. However, it should not be understood that the smallest particles give the highest desulphurization efficiency in fluidized bed combustors. For small particles the particle residence time in bed is short due to attrition and elutriation. There is then an optimum size depending on the medium size of coal ash in bed. The maximum degree of limestone conversion is equal about 50% (Hartman and Coughlin, 1974).

The efficiency of combustion gases desulphurization can be proximately described by a function

$$1 - \exp(-K(Ca/S)) \tag{4.106}$$

where Ca/S means a mole ratio of calcium fed into the bed to sulphur in coal and the value K depends on the type of coal, sorbent, temperature, and fluidization conditions within the bed. Figure 4.21 presents this function published by Highley (1975) for limestone. For a given temperature the fluidization velocity influences the desulphurization efficiency considerably. The higher is the velocity the smaller is the efficiency, mainly due to elutriation. The high values of efficiency above 90% could be obtained only for high excess calcium Ca/S > 3. At stechiometric ratio Ca/S = 1 the desulphurization does not exceed 50%.

Roberts et al. (1975) analyzed the influence of pressure on desulphurization for dolomite and limestone. Figure 4.22 presents the efficiency of atmospheric and pressurized (0.35–0.6MPa) combustors with bed temperature about 800°C. For limestone the desulphurization efficiency in the pressurized

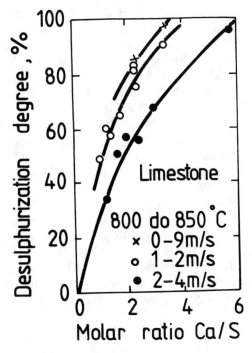

Fig. 4.21. Influence of Ca/S ratio on SO₂ reduction (Highley, 1975).

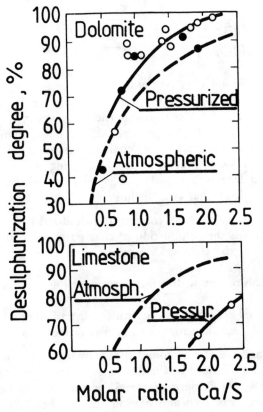

Fig. 4.22. Influence of Ca/S ratio on SO₂ reduction (Roberts et al., 1975): ○—coal, ●—heavy oil.

combustor dropped considerably. This could be expected because $CaCO_3$ does not calcinate in temperature 800°C at high partial pressure of CO_2 close to 0.07 MPa, so that binding of SO_2 is limited only to the outside layer of the limestone particle. For dolomite, however, pressure should not influence its reactivity because $MgCO_3$ calcinates at lower temperature forming large porosity particles. It has been demonstrated that higher pressure even improves the desulphurization of dolomite.

Attrition of sorbent particles within the bed increases elutriation. This is caused mainly by the fact that calcination changes the hard sorbent into material which is prone to attrition. However, attrition can have also a positive influence because crashing of the particles sulphated layer exposes fresh pores of highly reactive material.

Large ratios of Ca/S necessary to obtain satisfying desulphurization of flue gases would create transport cost of sorbent in industrial combustors. To achieve 90% of desulphurization during combustion of coal with 2.5% of sulphur it is necessary to use about 0.1–0.15 kg of dolomite per 1kg of coal. It is, however, possible to regenerate the sulpharized sorbent. The simplest way of $CaSO_4$ regeneration is through reaction with reducing gases at temperature of 1100°C according to mechanism

$$CaSO_4 + \begin{matrix} CO \\ H_2 \end{matrix} \rightarrow CaO + SO_2 + \begin{matrix} CO_2 \\ H_2O \end{matrix} \qquad (4.107)$$

High concentration of SO_2 in the outflowing gases does not allow production of sulphur as a byproduct of combustion. Regeneration should allow reduction of the sorbent amount by several times (Moss, 1975).

4.7. MATHEMATICAL MODELS OF COAL COMBUSTION IN FLUIDIZED BEDS

For the purpose of modelling of fluidized bed combustors it is essential to establish the mode of coal particle combustion. Most frequently a shrinking sphere mode is assumed for the char particle. However, this is often questioned. Recently Jung and La Nauze (1983a) have found that petroleum coke (initial diameter 7–13mm) and brown coal char (initial diameter 3.5–6.4mm) burn in a manner close to that of an ideal shrinking sphere.

Turnbull and Davidson (1984) published four important observations concerning the mode of char combustion in a fluidized bed:

- For the porosity of chars formed in a fluidized bed from medium or low-rank coals the combustion of char particles smaller than 2mm is controlled by the chemical reaction rate and the diffusion of oxygen in the pores.
- Combustion occurs within the pores near the external surface of the particle.
- The observed activation energy for fluidized bed combustion is about one half of the value from the intrinsic rate constant.

Jung and Stanmore (1980) found that brown coal particles of diameter 2–8.5mm burn in a fluidized bed under kinetic control with significant pore burning. Tomeczek and Remarczyk (1986) on the basis of experimental evidence for combustion of coal and partially devolatilized char in a fluidized bed reported not much difference in the respective mechanisms of combustion. The particles burn almost as shrinking spheres.

4.7.1. Modelling of char particle combustion

Let us consider a combustion of a batch of closely sized char particles injected into a hot air fluidized bed of inert particles. Assuming a two-phase fluidization theory the char particle reacts with oxygen within the emulsion phase. If the oxygen concentration within the emulsion is uniform, then for a shrinking sphere mode the rate of particle combustion is equal

$$\frac{d(d_p)}{dt} = \frac{24}{\rho_p} k \, C_{O_2,e} \tag{4.108}$$

where the rate constant can be described by equation (2.30) for an effective combustion reaction

$$C + O_2 \rightarrow CO_2 \tag{4.109}$$

The oxygen concentration within the emulsion can be found by means of oxygen balance:

rate of oxygen input into the bed $= w A_B C_{O_2,o}$ (4.110)

rate of oxygen leaving the emulsion $= w_{mf} A_B C_{O_2,e}$ (4.111)

rate of oxygen leaving the bed with the bubbles =

$$(w - w_{mf}) A_B \left[C_{O_2,e} + (C_{O_2,o} - C_{O_2,e}) \exp\left(-\frac{\dot{V} h_B}{w_a V_b}\right) \right] \tag{4.112}$$

where the effective volume flow $\dot{V}$ from a bubble to the emulsion phase, the absolute bubble rising velocity w_a, and the bubble volume V_b, can be calculated if a constant bubble diameter is assumed.

The rate of oxygen consumption within the bed is equal

$$(4.110) - ((4.111) + (4.112))$$

$$= (C_{O_2,o} - C_{O_2,e}) A_B \left[w - (w - w_{mf}) \exp\left(-\frac{\dot{V} h_B}{w_a V_b}\right) \right] \tag{4.113}$$

Because one mole of oxygen reacts with one mole of carbon, then the rate of oxygen consumption must be equal to $\frac{1}{12}$ of the rate of particle mass loss

$$(C_{O_2,o} - C_{O_2,e}) A_B \left[w - (w - w_{mf}) \exp\left(-\frac{\dot{V} h_B}{w_a V_b}\right) \right] = -\frac{m_{po}}{12} \frac{3 d_p^2}{d_{po}^3} \frac{d(d_p)}{dt} \tag{4.114}$$

where m_{po} is the initial mass of the char particle batch.

Eliminating oxygen concentration in emulsion phase by means of equation (4.108) Ross (1982) derived the following expression for the burn-out time of a mass m_{po} of char particles injected into an isothermal bed

$$t_c = \frac{m_{po}}{12 \, C_{O_2,o} A_B \left[w - (w - w_{mf}) \exp\left(-\frac{\dot{V} h_B}{w_a V_b}\right) \right]} + \frac{\rho_p d_{po}^2}{48 \, Sh \, D_{O_2} C_{O_2,o}} + \frac{\rho_p d_{po}}{24 \, k_C C_{O_2,o}} \tag{1.115}$$

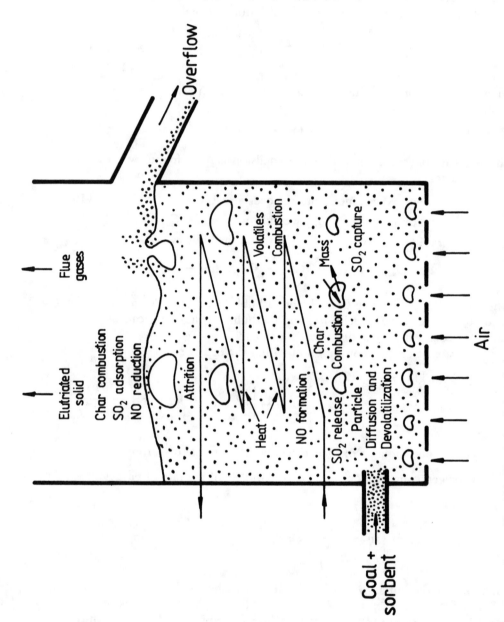

Fig. 4.23. Chemical and physical phenomena during coal combustion in a fluidized bed.

148

where the Sherwood number is equal $Sh = 2\,\varepsilon_{mf}$ for large char particles and $Sh = 2$ for particles smaller than the bed material.

It is important to notice that equation (4.115) can be used only in cases when the bed temperature is known. For large particles >3mm the combustion rate in fluidized beds is controlled by oxygen diffusion through the boundary layer of the particle and the time of combustion is a square function of particle initial diameter, while for small particles the second term is small and the combustion time is a linear particle diameter function.

4.7.2. Comprehensive model of coal combustion

The model presented in point 4.7.1. can not be used for design of practical combustors; however, it presents the mechanism of solid particle combustion in a two phase fluidized bed model. First of all the temperature of the bed cannot be assumed but has to be a subject of modelling. A model with practical combustors design capability should allow us to simulate the following parameters:

- combustion efficiency,
- char and limestone elutriation,
- overflow solids withdrawal,
- sulphur dioxide emission,
- nitrogen oxide emission,
- heat removel by cooling tubes,
- flue gas temperature,
- bed temperature.

Although many models of fluidized bed combustors were published, only a few of them have the possibilities to predict the performance of fluidized bed combustors under a wide range of operating conditions. The most comprehensive model with the broadest possibilities has been published by Rajan and Wen (1980). The model takes into account all the processes occurring in fluidized bed combustors described before in this chapter, and presented in figure 4.23. As the result of model testing for two pilot-scale combustors the authors found that an accurate estimation of model parameters is critical to make accurate predictions by the model.

4.8. AIR DISTRIBUTORS

The efficiency of heat and mass transfer between phases within a fluidized bed is determined by the way of distribution of combustion air supplied at the bottom of the combustor. There is a rich variety of air distributors applied in fluidized bed combustors (Basu, 1984; Werther, 1978) which can be generally divided into two groups:

- perforated plates,
- nozzle distributors.

The main duty of the distributor is to supply air uniformly into the fluidized bed. In most cases it supports also the weight of the bed material. An improperly designed distributor can lead to poor combustion efficiency and very often to coal particle agglomeration or even slagging. Two parameters are used to characterize the quality of distributors:

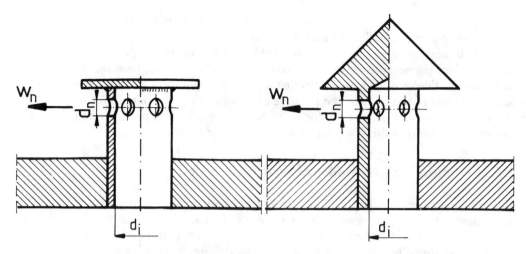

Fig. 4.24. Two types of nozzle distributors.

- dimensionless free cross-section, ratio of the air flow surface area to the surface area of the distributor,
- air pressure drop.

For distributors with large dimensionless free cross-section and low air pressure drop the flow of air will certainly not be uniform. The ratio of bed pressure drop Δp_B to distributor pressure drop Δp_d is used as a design parameter. For proximate considerations this ratio can be assumed to be not larger than one. Soroko et al. (1965) related this ratio to fluidization conditions within the bed

$$\frac{\Delta p_B}{\Delta p_d} \leqslant \frac{w^2 - w_{mf}^2}{w^2} \frac{1 - \varepsilon}{\varepsilon - \varepsilon_{mf}} \tag{4.116}$$

Qureshi and Creasy (1979) developed an equation for uniform fluidization condition

$$\frac{\Delta p_B}{\Delta p_d} \leqslant \frac{1}{0.01 + 0.2(1 - \exp(-d_B/h_{B,mf}))} \tag{4.117}$$

The optimum distributor pressure drop is still in dispute. Agarwal et al. (1962) specified the ratio $\dfrac{\Delta p_B}{\Delta p_d} \leqslant 10$, while Zenz (1968) recommends this ratio $\dfrac{\Delta p_B}{\Delta p_d} \leqslant 3$. To choose a proper pressure drop at the distributor it should be remembered that poor fluidization due to nonuniform air distribution can cause damages to the combustor, and in case of doubt a high rather than a small value of distributor pressure drop ought to be taken.

4.8.1. Nozzle distributors

In most nozzle distributors air flows through a number of horizontal or slightly downwards tilted nozzles. Figure 4.24 presents two types of nozzles. Interaction of air jets flowing out of the neighbouring nozzles cause formation of large air bubbles. To prevent this it is advised to arrange the nozzles in

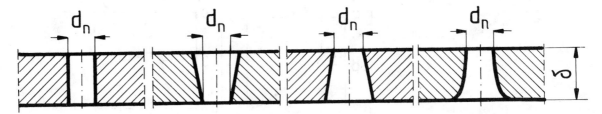

Fig. 4.25. Examples of perforated distributors.

different fashion on the whole distributor plate. A tapered roof helps to diminish the formation of dead zones.

Because it is possible to place the nozzles at a certain height above the solid material supporting plate, then there is always a layer of inert solid particles on the plate insulating it thermally and preventing its distortion. One of the advantages of nozzle distributors is the near elimination of the backflow of solid particles.

The pressure drop at the distributor nozzle can be calculated by equation

$$\Delta p_d = \xi_d \frac{\rho_g w_n^2}{2} \qquad (4.118)$$

where for: $\Theta_d = 0.03$–0.21, $\dfrac{d_n}{d_i} = 0.153$–0.4 and $Re = \dfrac{w_n d_n}{\nu_g} = 5600$–$15800$, the discharge number is equal (Tomeczek et al.,1990)

$$\xi_d = 134 \left(\frac{d_n}{d_i}\right)^{-0.17} Re^{-0.1} \Theta_d^{0.66} \qquad (4.119)$$

The distance between the nozzles on the plate of the distributor is an important design parameter and it can be related to the length of horizontal penetration of the jet flowing out a nozzle orifice. Merry (1971) published an equation for the length of penetration

$$\frac{L}{d_n} = 5.25 \left[\frac{\rho_g w_n^2}{(1 - \varepsilon)\rho_p d_p g}\right]^{0.4} \left(\frac{\rho_g}{\rho_p}\right)^{0.2} \left(\frac{d_p}{d_n}\right)^{0.2} - 4.5 \qquad (4.120)$$

4.8.2. Perforated distributors

The simplest type of distributor form is a perforated plate with vertical holes which can be straight, convergent, or divergent as presented in figure 4.25. Perforated plates have two main disadvantages:

- possibility of dead zones formation on the plate supporting the bed material,
- backflow of solid particles through the holes.

The distributor pressure drop can be calculated by equation (4.118) with the discharge number for a perforated plate equal (Hydhmark and O'Conell, 1957)

$$\xi_d = \frac{0.102(1 - \Theta_d^2)}{c^2} \qquad (4.121)$$

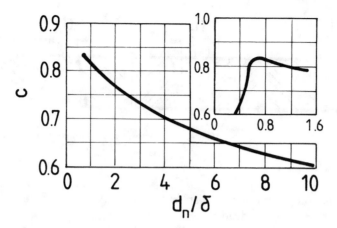

Fig. 4.26. Discharge number of a perforated distributor (Oroczko et al., 1959).

where the value c is related to the distributor thickness δ

$$c = 0.835 \left(\frac{d_n}{\delta}\right)^{-0.133} \tag{4.122}$$

For small (d_n/δ) the value c should be taken from figure 4.26.

The height of vertical jet penetration into a fluidized bed can be calculated by equation (Merry, 1975)

$$\frac{h_{bo}}{d_n} = 5.2 \left(\frac{\rho_g\, d_n}{\rho_p\, d_p}\right)^{0.3} \left[1.3 \left(\frac{w_n^2}{g\, d_n}\right)^{0.2} - 1\right] \tag{4.123}$$

To avoid the harmful dynamic impact of the jet on the cooling tubes submerged in the bed it is advised to place the tubes at a height above the value given by equation (4.123).

To prevent too early formation of bubbles due to collisions of jets flowing out of holes, the minimum distance between the holes should be equal to 1.5 d_{bo}. For industrial scale distributors the distance between the holes of the distributor can be large, some several milimeters. The regions between the holes can form dead zones of nonfluidized material presented in figure 4.27. The largest dead zones can be formed in the bed when a stream of air insufficient for fluidization flows through some of the holes. These holes do not function then; however they are not blocked. For bed fluidization number $\frac{w}{w_{mf}} \approx 1$ a considerable part of the holes can be out of operation. It has been stated also that in some cases a fluidization number over 1.4 was necessary to obtain proper flow of air through all the holes of the perforated plate. For higher beds it is easier to get a functioning of all holes at small fluidization numbers. Increase of distributor pressure drop helps to obtain uniform fluidization without dead zones.

The backflow of solid particles through the holes of a perforated plate cannot be avoided, even at very high air pressure drop through the distributor $\frac{\Delta p_B}{\Delta p_d} = 0.8$. The stream of particles flowing back through the distributor decreases considerably with increasing thickness of the distributor plate. Briens and Bergougnou (1985) found that the backflow of solids can be reduced

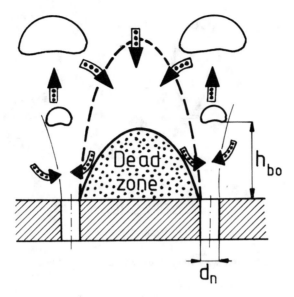

Fig. 4.27. Mixing of solids above perforated distributor.

over 100 times as a result of a plate thickness increase from 3cm to 10cm. Wen et al. (1980) proposed that the orifice diameter should not be larger than 3 to 8 times the particle diameter.

4.9. FLUIDIZED BED COMBUSTORS

One of the main advantages of fluidized bed combustors is the possibility to burn a broad variety of coals, even wastes. The geometrical characteristics of the combustor depend on the type of coal and the size of coal particles. Among the variety of developed designs, there are two general kinds of combustors:

- without submerged heat exchanger,
- with submerged heat exchanger.

From the point of fluidization conditions we can divide also the combustors into two groups:

- bubbling beds,
- circulating beds.

Coal feeding should organize a uniform fuel distribution throughout the bed. There are three main ways of coal feeding presented in figure 4.28. Pneumatic coal supply can be applied to dry coals (moisture below 12%). In order to get proper coal mixing with the inert bed material, typically one feeding point is required per one square metre of the combustor cross-section. Coal is fed laterally with velocity 20m/s just above the air distributor. Since coal is introduced to the bottom part of the bed it creates good conditions for combustion of the volatiles released from coal within the bed.

Overbed feeding has the advantage of simplicity of the equipment. This system is very convenient for large beds where the complicated network of pneumatic feeding pipes would be very expensive. One spread feeder can cover

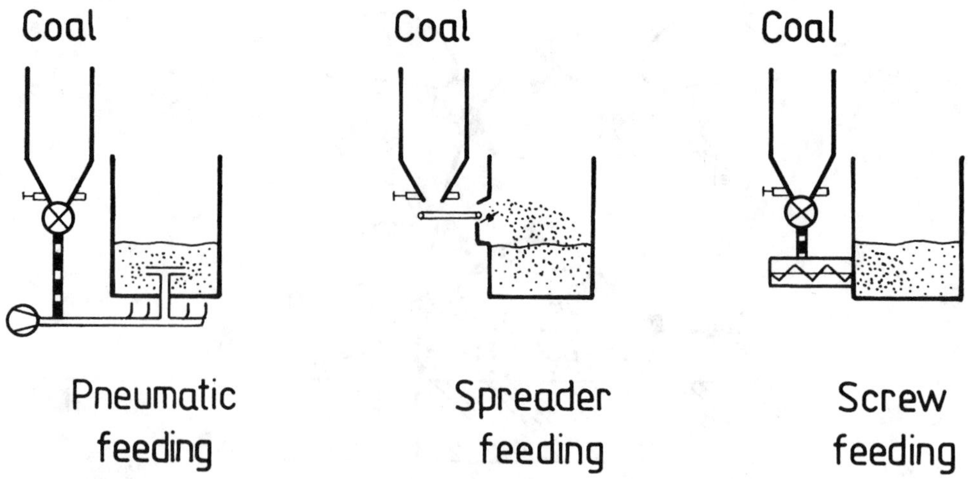

Coal   Coal   Coal

Pneumatic   Spreader   Screw
feeding   feeding   feeding

Fig. 4.28. Three ways of coal feeding into fluidized bed.

a bed surface of about 3m². The main disadvantage of the overbed spread feeding is caused by segregation of particles by the upflowing gases and the devolatilization of coal on the bed surface. This can be overcome through partial separation of fines from coal before feeding. The coarse particles can then be introduced by the overbed spreader while the fines are introduced by the pneumatic feeder to the bottom part of the bed.

Screw feeding can be realized both into the bed and over the bed. Some fuels show the tendency to form coarse lumps at the discharge of screw feeder. To prevent it an air jet underneath the discharge hole can be used to disperse the coal particles. To prevent devolatilization and agglomeration of coal along the screw feeder a water cooling of the feeder jacket can be applied. This system has been successfully applied also in pressurized beds up to several bars.

Sorbent can be mixed with coal before feeding into the combustor, but most frequently a separate feeding device directly into the bed is used.

Solid material in bed contains mainly a mixture of partly sulphated sorbent and coal ash. Only about 1% of bed mass makes carbon. Withdraw of bed material can be done by overflow or through ash ports placed on the distributor plate. High temperature ash is cooled mostly in a separate fluidized bed heat exchanger by incoming combustion air from bed temperature 850°C to about 200°C. Figure 4.29 presents a placement of the ash removal ports together with the pneumatic coal feeding nozzles and air nozzles on the bed material supporting plate. The bed height is kept constant at full load by removing the bed ash at a controlled rate. In the case of coals with low mineral matter content it is necessary to supply additional inert material to build up the bed height.

Fine ash captured in cyclone is recycled to the bed together with coal to increase the coal burn-out.

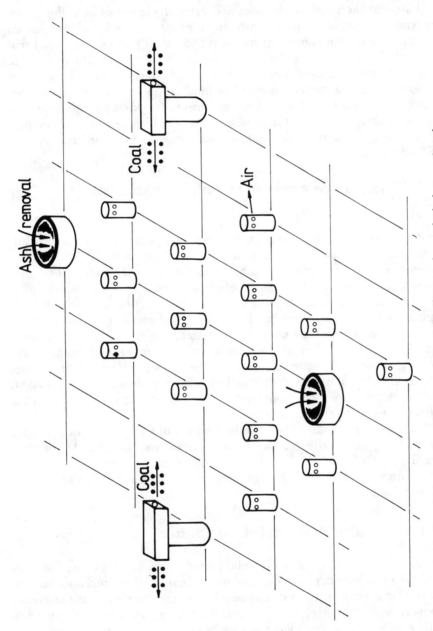

Fig. 4.29. Ash removal ports and the coal feeding nozzles on the bed supporting plate.

4.9.1. Fluidized bed combustors without immersed heat exchangers

Combustors without heat exchangers generate only high temperature flue gases. To keep the bed temperature on the level 800–950°C it is necessary to remove the excess heat by gases. Mostly this is done by supplying additional air, but the same can be obtained by recirculation of cold flue gases. This type of combustor is particularly useful in case of waste fuel combustion when the low carbon content in fuel is sufficient only to heat the solid material and the near-stechiometric air to the bed temperature. The absence of cooling tubes in bed does allow application of coarser coal, up to several milimetres, and as a consequence higher fluidization velocity. A tapered bed is particularly useful for this kind of combustor. The lowest heating value of fuel can be calculated from the condition that the liberated heat of combustion will heat up the fuel solid and the stechiometric air to bed temperature of 850–950°C.

4.9.2. Fluidized bed combustors with immersed heat exchangers

Combustors with immersed heat exchangers can operate at lower excess air. For general coal we can assume that about 50–60% of combustion heat must be removed from bed by cooling tubes, and the rest will be utilized by convection heat exchangers from the flue gases. Experiments have shown that corrosion and erosion of tubes submerged in bubbling beds is not a serious problem. Within the bed tube bundle, steam is generated. At full load all the tubes are immersed in the fluidized bed. Decreasing the bed height through removal of bed material to ash containers, we free part of the tubes from the bed. That causes a decrease of steam generation due to lower heat transfer in the freeboard compared to that in bed. The variation of the bed height, by filling the bed with ash from containers or removing it from the bed, is commonly used as an effective way of load control.

Starting up of fluidized bed combustors is done by some oil or gaseous burner heating the inert bed either from above or from bottom. At temperatures about 750°C coal is injected and the oil or gas burners shut off. Figure 4.30 presents an example of a combustor with a bubbling bed, also called a stationary bed.

4.9.3. Circulating fluidized bed combustors

Limited rate of lateral diffusion of solid particles within a bubbling fluidized bed needs a high number of feeding points to obtain uniform fuel distribution. Also the large cross-sections of industrial scale combustors limit the power of boilers with stationary beds. Increasing the gas velocity to some 8m/s together with reduction of solid particles diameter to 1mm, we cause the bed expansion and decrease of its density. Solid particles move upwards with velocity much slower than that of gas. Figure 4.31 presents qualitatively the gas-solid particles velocity difference for combustion chamber with increasing velocity. In stationary fluidized beds the relative gas-solid velocity is small. For higher velocities the solid particles move upwards much faster. That causes intensive elutration from bed. A cyclone placed at the combustion

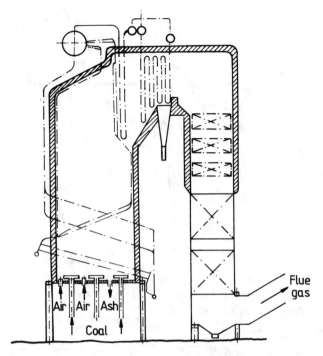

Fig. 4.30. A fluidized bed combustor with a bubbling bed.

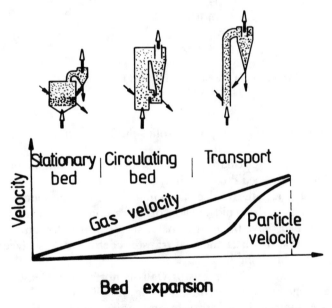

Fig. 4.31. Velocity of gas and of solids in combustion chambers of
three characteristic combustion types.

chamber outflow captures the solids and recycles them back to the bed, hence
the name of this kind of combustor. In circulating beds the velocity of particles is higher than in stationary beds, but at the same time the relative gas-
particle velocity is much higher, hence very intensive heat and mass transfer
occurs between the solid and gaseous phases. Increasing further the gas ve-

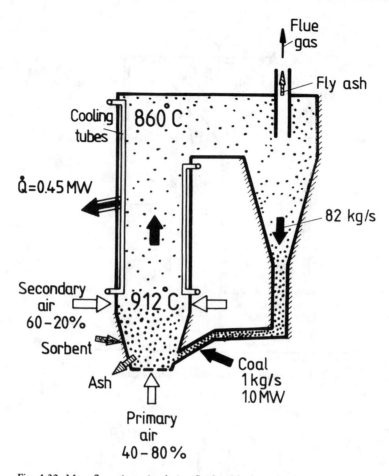

Fig. 4.32. Mass flows in a circulating fluidized bed combustor.

locity we get a pneumatic transport of solids during which the relative velocity decreases again.

Combustors with circulating beds do allow us to obtain much higher power per 1m² of the bed cross-section, but on the other hand they must be almost as tall as the pulverized coal combustors to accommodate sufficient heat transfer surface. To avoid erosion of tubes immersed in beds the circulating fluidized beds have the heat transfer surfaces installed on the chamber walls in the freeboard, where the heat transfer coefficient to surfaces is lower than those of the bubbling bed.

The lower 25% of the combustor walls is not cooled. Removal of tubes from the bed together with intensive mixing of solids does allow drastic reduction of the number of necessary feeding points in comparison with bubbling beds.

The combustion air is supplied at multi points. The primary air flowing through the distributor consists 40–60% of the combustion air, while the rest flows through secondary air nozzles placed on all four walls at various levels (Fig.4.32). Intensive circulation of solid particles up to 100 times the coal mass stream fed into the combustor causes a proper removal of heat generated in the bed to the upper cooling walls. As a consequence, the temperature difference within the combustor between the bed and the freeboard

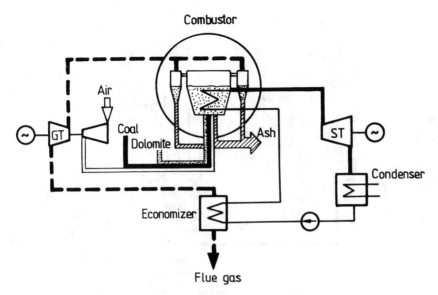

Fig. 4.33. A pressurized combustor coupled with steam and gas turbines.

is minimized. That creates favourable conditions for char particle combustion and an environment suitable for sulphur capture by the sorbent. Ash removed from the combustor is cooled in a separate fluidized bed heat exchanger. The main advantages of circulating fluidized bed combustors are:

- high coal burn-out, 99%,
- high fuel throughput per one feeder, 150 MW$_{th}$ per one feeding point,
- minimized emission of sulphur dioxide, nitrogen oxides, and hologens.

The main disadvantage of CFBC is caused by higher air pressure necessary for solids circulation. Because the cyclone separator most frequently is not cooled, the ceramic walls also reduce the rate of start up from the cold stage.

4.9.4. Pressurized fluidized bed combustors

The advantages of fluidized bed combustion are stronger at elevated pressure. Pressurized combustors generate steam and high pressure hot gaseous combustion products. Combustion of coal in fluidized bed of pressure about 1.2 MPa does allow for much higher volumetric loads. Much larger heights of stationary bed combustors (3.5m) lead to better coal burn-out (over 99%), even without fines recirculation. In order to remove intensive heat fluxes from the bed it is necessary to apply placement of immersed tubes more densely than in atmospheric beds.

Figure 4.33 presents a typical flow diagram of a pressurized combustor coupled with steam and gas turbines. The air at pressure 1.2MPa is supplied to a fluidized bed combustor. The majority of combustion heat is transferred to the immersed tubes. Generated steam is used to drive the steam turbine. The hot combustion gases are cleaned in cyclones to remove ash and sorbent particles, and then expand in the gas turbine. The enthalpy of the flue gases leaving the gas turbine is utilized by the feed water. The combustor is enclosed

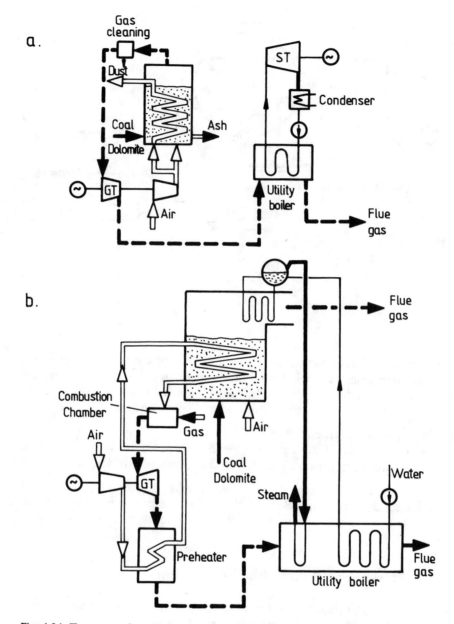

Fig. 4.34. Two ways of particulates concentration reduction in flue gases for gas turbine driving.

in a pressure vessel cooled by the incoming air. The combined gas-steam cycle driven by a pressurized coal fired combustor in which the gas turbine produces about 20% of power can reach overall net efficiency up to 41%.

The cleaning of hot gas particulate in cyclones before the gas turbine and at the back-end by filter before the chimney, is crucial to meet turbine material and environmental demands. It is possible to diminish the danger of turbine blades erosion by reduction of the particulates concentration in hot gases in a way presented in figure 4.34. The tubes immersed in the bed are cooled by pressurized air, which leaving the bed at temperature about 850°C is used to drive the gas turbine. In case "a" the particulates concentration is considerably reduced, while in "b" it is eliminated but additional fluid fuel must be used to increase the efficiency of power production.

REFERENCES TO CHAPTER 4

Agarwal J. C., Davis W. L. and King D. T.—Chem. Engng. Prog., 58(1962)85.

Ainstein V. G. in Zabrodsky S. S.—Hydrodynamics and Heat Transfer in Fluidized Beds. MIT Press, Cambridge, Massachusetts, 1966.

Andeen B. R. and Glicksman L. R.—ASME-AICHE Heat Transfer Conference, Paper No 76-HT-67, St. Louis, MO, August 9–11(1976).

Babu S. P., Shah B. and Talwalker A.—AIChE Symposium Series, No 176, Vol. 74(1978).

Baerg A., Koassen J. and Gischler P. E.—Can. J. Research 28F (1950)287–307.

Baskakov A. P.—Izd. AN USSR., Energ. and Transport, 3(1966)122.

Baskakov A. P., Berg B. V., Vitt O. K., Filipovsky N. F., Kirakosyan V. A., Galdobin J. H. and Maskaev V. K.—Powder Technology, 8(1973)273–282.

Basu P. in Basu P.(ed.)—Fluidized Bed Boilers: Design and Application. Pergamon Press, Toronto, 1984.

Bellgardt D.—Ph.D. Dissertation, Technical University of Hamburg, 1985.

Bennet C. O. and Myers J. E.—Momentum, Heat and Mass Transfer. McGraw-Hill Company, New York, 1962.

Bock H. J. and Schweinzer J.—Chem. Ing. Tech., 57(1985)486–487.

Borgwardt R. H.—Environ. Sci. Tech., 41(1970)59.

Borgwardt R. H. and Harvey R. D.—Environ. Sci. Tech., 6(1972)350.

Borodulya V. A., Epanow J. G. and Teplickij J. S.—Ing Fiz. Jurnal, 5(1982)767.

Borodulya V. A., Ganzha V. L. and Podberezsky A. I. in Grace J. R. and Matsen J. M. (eds.)—Fluidization. Plenum Press, New York, 1980.

Botterill J. S. M.—Fluid-bed heat transfer. Academic Press, London, 1975.

Botterill J. S. M. and Sealey C. J.—British Chem. Engineering, 15(1970)9, 1167–1168.

Botterill J. S. M. and Williams J. R.—Trans. Inst. Chem. Engrs, 41(1963)217.

Briens C. L. and Bergougnou M. A.—Powder Technology, 43(1985)235–241.

Brotz W.—Podstawy Inzynierii Reakcji Chemicznych. WNT, Warszawa, 1964.

Carman P. C.—Flow of gases through porous media. Academic Press Inc., New York, 1956.

Chen J. L. P. and Keairns D. L.—Can. J. Chem. Engn., 53(1975)395.

Chmielewski A. and Selecki A.—Ing. Chem., 3(1977)549.

Ciborowski J.—Przegl. Chem., 6(1948)164–172.

Ciborowski J.—Fluidyzacja. PWT, Warszawa, 1957.

Darton R. C.—Trans. Inst. Chem. Engng., 2(1979)134.

Davidson J. F. and Harrison D.—Fluidized Particles. Cambridge Univ. Press, England, 1963.

Davidson J. F. and Harrison D.—Fluidization, Academic Press, London, 1971.

Emielianov I. D., Mieszczeriakov W. D. and Slinko M. G.—Chim. Prom., 6(1967)15.

Fan L. T., Song J. C. and Yutani N.—Chem. Eng. Sci., 1(1986)117.

Gibbs B. M. and Hampartsoumian E. in Basu P. (ed.)—Fluidized Bed Boilers: Design and Application. Pergamon Press, Toronto, 1984.

Gibbs B. M., Pereira F. J. and Beer J. M.—16th Symp. (Int.) on Combustion. The Comb. Inst., Pittsburgh, 1976.

Gelpierin N. I., Ainstein V. G. and Zaykovskii A. V.—Chim. Prom., 6(1966)418.

Gelpierin N. I., Kruglikov V. Ya. and Ainstein V. G.—in Ainstein V. G. and Gelpierin N. I.—Int. Chem. Engng., 6(1966) 67–73.

Gelpierin N. I., Ainstein V. G. and Korotjanskaya L. A.—Chim. Prom., 6(1968)427.a.

Gelpierin N. I., Ainstein V. G. and Zaykovskii A. V.—Chim. i Nieft. Masch., 3(1968)17. b.

Grewal N. S. and Saxena S. C.—Int. J. Heat Mass Transfer, 23(1980)1505–1519.

Grewal N. S. and Saxena S. C.—Ind. Eng. Process Des. Dev., 20(1981)108.

Grewal N. S. and Saxena S. C.—Ind. Eng. Chem. Process Des. Dev., 22(1983)367–376.

Hartman M. and Coughlin R. W.—Ind. Eng. Chem. Process Des. Dev., 13(1974)248.

Highley J.—Institute of Fuel Symposium Series. No 1: Fluidized Combustion, 1975.

Highley J. and Merrick D.—AICHE Symp. Ser., No 116, Vol. 67(1971)219.

Hydhmark G. A. and O'Connell H. E.—Chem. Eng. Progr., 53(1957)127.

Jung K. and La Nauze R. D.—Can. J. Chem. Eng., 61(1983)262–264, a.

Jung K. and La Nauze R. D. in Kunii D. and Tei R. (eds.)—Proc. Fourth Int. Conf. in Fluidization. 29.05-3.06.1983. Koshi Kojima, b.

Jung K. and Stanmore B. R.—Fuel 59, (1980)74–80.

Karchenko N. V. and Makhorin K. E.—Inst. Chem. Eng., 4(1964)650–654.

Kirner W. in Lowry H. H. (ed.)—Chemistry of Coal Utilization. John Wiley and Sons, New York, 1945.

Kobayashi M., Arai F. and Sunakava T.—Kagaku Kogaku, 31(1967)239.

Kono M.—AIChE, 72nd Annual Meeting, San Francisco, Ca, Nov. 25–29, 1979.

Kozeny J.—Ber. Wien Akad., 136a(1927)271.

Kudzia W.—PhD Thesis. Silesian Technical University, 1980.

Kunii D. and Levenspiel O.—Fluidization Engineering, John Wiley, N. York, 1969, a.

Kunii D. and Levenspiel O.—Journal of Chem. Eng. of Japan 1(1969)122, b.

Large J. F., Martine Y. and Bergougnou M. A. in Yates J. G. (ed.)—Fundamentals of fluidized bed chemical processes. Butterworths, London, 1983.

Leva M., Grummer M., Weintraub M. and Polchik M.—Chem. Eng. Progr., 44(1948)511–520.

Lewis W. K., Gilliland E. R. and Bauer W. C.—Ind. Eng. Chem., 41(1949)1104–1117.

Locke H. B., Lunn H. G., Hoy H. R. and Roberts A. G.—Fourth Int. Conf. in Fluidized Bed Combustion. MITRE Corp., Mc Lean, Virginia, 1975.

Lorkiewicz Z. and Jastrzab Z.—Zesz. Nauk. Pol. Sl., Energetyka, Z. 79(1982)113–122.

Martin H.—Chem. Eng. Proces., 18(1984)157–169.

Mc Lahren J. and Williams D. F.—J. Inst. Fuel, 42(1969)303–308.

Merrick D. and Highley J.—AICHE Symposium on "Control of particulate emissions from gaseous fluidized beds". New York, Nov. 1972.

Merry J. M. D.—Trans. Instn. Chem. Eng., 49(1971)189.

Merry J. M. D.—AICHE J., 21(1975)507.

Mickley H. S. and Fairbanks D. F.—AICHE J., 1(1955)374–384.

Miller C. D. and Longwinuk A. K.—Ind. Eng. Chem., 43(1951)1220–1225.

Mori S. and Wen C. Y.—AICHE J., 21(1975)109.

Moss G.—Institute of Fuel Symposium Series No 1: Fluidized Combustion, 1975.

Newby R. A., Vaux W. G. and Keairns D. L. in Kunii D. and Tei R. (eds.)—Proc. Fourth Int. Conf. in Fluidization. 29.05.-3.06.1983, Koshi Kojima.

Oroczko D. I., Basov V. A. and Melik-Ahnazarov T. H.—Chim. Techn. Topliv i Masiel, 4(1959)54.

Pattipati R. R. and Wen C. Y.—Ind. Eng. Chem. Proc. Des. Dev., 21(1982)784–786.

Pereira F. J., Beer J. M., Gibbs B. and Hedley A. B.—15th Symp. (Int.) on Combustion. The Combustion Inst., Pittsburgh, 1974.

Petrie J. C., Freeby W. A. and Buckham J. A.—Chem. Eng. Prog., 64(1968)45–51.

Podkowa K.—Koks, Smola, Gaz, 7–8 (1970)210–214.

Quereshi A. E. and Creasy D. E.—Powder Technology, (1979)113–119.

Rajan R. R. and Wen C. Y.—AICHE J., 26(1980)642–655.

Rajan R. R., Krishnan R. and Wen C. Y.—AICHE Symposium Ser. No 176, 74(1978)1123.

Roberts A. G., Stantan J. E., Wilkins D. M., Beacham B. and Hoy H. R.—Institute of Fuel Symp. Ser. No 1: Fluidized Combustion, 1975.

Ross I. B. and Davidson J. F.—Trans. Inst. Chem. Engrs., 60(1982)108.

Rowe P. N., Nienow S. W. and Aglium S. J.—Trans. Inst. Chem. Engrs., 50(1972)310a, 324b.

Shannon P. T.—PhD Thesis, IIT, Chicago, Illinois, 1959.

Shelef M. and Kummer J. T.—Chem. Eng. Prog. Symp. Ser., 67(1971)74.

Shirai T.—PhD Thesis, Tokyo Institute of Technology, 1954.

Sit S. P. and Grace I. R.—Chem. Eng. Sci., 36(1981)327.

Sitnai O.—Ind. Eng. Chem. Proc. Des. Dev., 20(1981)533.

Soroko B. E., Muhlenov I. P. and Hiljalev M. F.—Chimiya i Chimicheskaya Technologia, 4(1965)668.

Szekely J. and Fisher R. J.—Chem. Eng. Sci., 24(1969)833–849.

Ternovskaya A. N. and Korenberg Yu.G.—Pyrite Kilning in a Fluidized Bed. Izd. Chimiya, Moscow, 1971.

Thonglimp V., Hiquily N. and Laguerie C.—Powder Technology, 38(1984)233–235.

Tomeczek J. and Remarczyk L.—Can. J. Chem. Eng., 64(1986)871–874.

Tomeczek J., Jastrząb Z. and Gradoń B.—Powder Technology. In print (1991).

Tomeczek J., Jastrząb Z. and Gradoń B.—Report R-991/RM-4/89 of the Inst. of Fuel Energy, Silesian Techn. Univ., Katowice, 1990.

Turnbull E. and Davidson J. F.—AICHE J., 30(1984)881–889.

Valenzuela J. A. and Glicksman L. R.—Powder Technology, 38(1984)63.

Vreedenberg H. A.—Chem. Eng. Sci., 9(1958)52–60.

Wen C. Y. and Chen L. H.—AICHE J., 28(1982)117.

Wen C. Y. and Ishida M.—Environ. Sci. Tech., 1(1973)103.

Wen C. Y. and Yu Y. H.—Chem. Eng. Progr. Symp. Ser., 62(1966)100–111.

Wen C. Y., Chitester D. C., Kornosky R. M. and Keairns D. L.—AICHE J., 31, 7(1985)1086.

Wen C. Y., King D. F. and Shang J.—Proc. Conf. on Fluidized Bed Combustion, System Design and Operation. Morgantown, 1980.

Werther J.—Ger. Chem. Eng., 1(1978)166–174.

Wicke E. and Fatting F.—Chem. Ing. Techn., 26(1954)301–309.

Yagi S. and Aochi T. in Yates J. G. (ed.)—Fundamentals of fluidized-bed chemical processes. Butterworths, London, 1983.

Yagi S. and Kunii D.—AICHE J., 3(1967)373.

Yan-fu Shi and Fan L. T.—Powder Technology, 48(1985)23.

Yoshida K., Kunii D. and Levenspiel O.—Int. J. Heat Mass Transfer, 12(1969)529.

Yoshida K., Ueno T. and Kunii D.—Chem. Eng. Sci., 29(1974)77–82.

Zabrodsky S. S.—Hydrodynamics and Heat Transfer in Fluidized Beds. MIT Press, 1966.

Zeng F. A. and Weil N. A.—AICHE J., 4(1958)472.

Zenz F. A.—Proc. Tripartite Conf. Fluidization, 136, Montreal, 1968.

Zheng, Zhao-Kiang, Yamazaki R. and Jimbo G.—Kagaku Kogaku Ronbunshu, 11(1985)115–117.

SUBJECT INDEX